Very special greetings to Cathy
Felton and a warm welcome
to my flying circus.

Ernst H. Gann

1986

Ernest K. Gann's Flying Circus

by Ernest K. Gann

Ernest K.Gann's Flying Circus

Ernest K. Gann

Paintings by Robert Parks

Macmillan Publishing Co., Inc.,
NEW YORK

Macmillan Publishing Co., Inc.
866 Third Avenue, New York, N. Y. 10022
Collier-Macmillan Canada Ltd.

All portions of this book except "Old Number One,"
"Logan," "The Clippers," "The Little
Ones," and "Dossier on Principal Characters"
appeared previously in *Flying* magazine.
Copyright © 1968, 1972, 1973, and 1974
Ziff-Davis Publishing Company.
"Old Number One" appeared previously in
Private Pilot. Copyright © 1970 Macro-Comm
Corporation.

First Printing 1974

Printed in the United States of America

Library of Congress Cataloging in Publication Data

Gann, Ernest Kellogg, 1910–
    Ernest K. Gann's Flying circus.

    1. Aeronautics—History.   I.   Title: Flying circus.
TL515.G25   629.13'09'042   74-11050
ISBN 0-02-542400-9

Manned flight has done more
to change the relations of the earth's inhabitants
than any other factor of human endeavor.
As an awesome power for good and for evil
it has nullified the protection and isolation
of traditional frontiers.
It has lofted the common man toward hitherto
undreamed of destinations, and
it has changed forever,
the very pattern of his daily existence.

E.K.G.

Contents

List of Paintings

Preface

One midnight. In a time when most aircraft engines were round and reciprocating. Four of them are now hunched on our wings, snoring through the night.

I turn to look over my shoulder at our navigator, whose face is dimly outlined in the light reflected from his chart. He is watching a fly march across the spoke marks of his last celestial fix, frowning at the insect's migration, his eyes disapproving of its aimless course across a paper ocean.

"Where are we?"

The navigator shrugs his shoulders.

"It beats me. So I suppose we're lost."

He smiles. He is a very skillful navigator, and we know each other well because we have spent hundreds of hours together on various flight decks. We are over the South Atlantic, bound from Brazil to Ascension Island, which is barely a pimple in the liquid immensity below. Our navigator knows almost exactly where we are, or were a few minutes ago, and he knows I know he knows. Else he would have warned me. It is his momentary amusement to pretend ignorance of our position—a little game we play when the night is benign and all is going well.

Lost?

I turn away from his lingering smile and look out upon the wilderness of the night. Lost? Perhaps the navigator of this craft is not mocking me after all. For everywhere below is the hostile ocean, blacker than the night. And from this flight deck there is no visible measuring of its depth or dimensions. Everywhere above are glittering stars quite as unmeasurable as the void below. Then it could be that we are lost, or at least so detached from the real world that all we have known, including our very aging, has temporarily ceased to be.

For the next several hours we could be space travelers hurtling through an infinite vacuum, truant from our ordinary lives. There is nothing aloft or alow to indicate movement. Nothing.

For how long and how many times have I been spellbound thus? Ten years, a little less. And still always I yearn for more of life aloft, a continuance of this marvelous removal. Ten years and 6,000 hours are not enough. There are no new prophetic powers given me by the night. I cannot foresee that after another score of years and yet another eight, and hours beyond 17,000, I will still hold an insatiable lust for the subtle charms of flight.

Once the sensual pleasure of controlling an aircraft is mastered, exhilaration diminishes and incurable habit takes command. Once the secretly nursed conviction that something dangerous and dramatic might happen —with yourself as hero or victim—is forgotten, then the heart joins the mind in submission to the necromancy of flight. Pilots who have flown for fifty years are unable to shake it—whatever it is.

Perhaps it is the same for all pilots. As on that South Atlantic night, when I brooded about such things, the same removal can be known repeatedly to everyone who has passed first solo.

Are we lost, or are we found at last? On earth we strive for our various needs, because so goes the fundamental law of man. Aloft, at least for a little while, the needs disappear. Likewise, the striving.

In the thoughts of man aloft, good and evil become mixed and sometimes reversed. This is the open door to wisdom.

Aloft, the earth is ancient and man is young, regardless of his numbers, for there, aloft, he may reaffirm his suspicions that he may not be so very much. This is the gateway to humility.

And yet, aloft, there are moments when a man can ask himself, "What am I, this creature so important to me? Who is it rules me from birth to tomb? Am I but a slave destined to crawl from labor to hearth and back again? Am I but one of the living dead, or my own god set free?" This is the invitation to full life.

To look down frequently from high position and inspect what men have wrought is to create a stinging sense of shame in all who are not blind. This is the embrace of hope.

Yet from aloft, what men do to each other or to themselves remains concealed. So hatred does not insult the eye, and defeat becomes invisible. This is the restoration of faith.

Flight is not thrilling, nor should its unique pleasures be compared with the voluptuousness of sex, the bawdy joys of drink, or the dubious dreams of opiates.

"Where are we?"

"If you really must know, I'll tell you."

"Never mind. Here aloft, we are not lost, but found."

ERNEST K. GANN
San Juan Island

Ernest K. Gann's Flying Circus

Chapter One

Logan

One dawning. In a time before people spoke in aeronyms, except for some airmen who snarlingly referred to the new Civil Aeronautics Authority as the CAA, it had never occurred to any of us that its ancestor, the Department of Commerce, previously responsible for the licensing of ourselves and our craft, should be called D.O.C.

I have logged nearly four hundred hours air time (hardly more than fifty of which is "padded"), possess a "Limited Commercial" license and am working on my "Transport." When I achieve it that will be a gala day —the true mark of a professional. I have also applied to the U.S. Army Air Corps for selection as a flight cadet. It seems the only way to get on in this business and bag the minimum of a thousand hours the air lines insist you have before hiring. I have passed my physical and am awaiting word from the Air Corps. Those who are supposed to know say I may wait a long time —like forever, because the cadet programs are being cut back drastically. And I am acutely aware that on the assembly lines of two local aircraft plants there are pilots with thousands of hours, pounding rivets.

Now at dawn my youthful zest for flight is tremendous, compounding itself as I rise from darkness until every trifle becomes a vivid reflection of the whole experience. It is 1936: my beliefs in the purpose of man and in the order of things are still untarnished and I hardly know the sensation of fear.

At dawn there is still dew on the fuselage and it slithers daintily along the cowling which encircles me. The air is moist and cool, and so the airplane climbs superbly. From the moment of liftoff the ascent has been ebony smooth. Once the earth drops away I seem to be floating rather than flying, and in spite of the noise there is an uncanny sensation of silence. I am pleased at how the sound of an engine at dawn seems to be much less raucous than it is at other times, and yet the same sense of power is there ahead of me. Both the even rumble of the engine and the faint whopping of the propeller are muffled in the heavy air.

I find the sky above is lashed with long sweeping mares' tails already bronze and turning to gold. I watch the momentary purpling of the vague horizon to the west, then turn back to observe the upward chase of the very first eastern fire. I glance down at the sullen brown-blue haze which is still the night coverlet of earth, then bank steeply and hold in a spiral. It is only necessary to shift my eyes for a view either below or aloft.

Up here there is pure solitude. Up higher still, I fancy that I might feel exalted as a god, an innocent vision which I smilingly reject as too ethereal for a man whose flying is merely his job. I am not a poet, but a solitary worker assigned to the air. Here at ten thousand feet, with the cold slip-stream numbing my cheeks, here in the dawn while it seems that all of my fellow men are still asleep, it is easier to think and to better assess my situation.

I am twenty-six years old and consider myself regularly employed in the flying profession, a considerable gilding of fact. My brown leather jacket is polished with wear. I have slept in it, worked in it, loved in it and lived in it, until it has become like a second skin. My leather helmet is much newer, still so tight it binds my forehead. Now leveling out, I push my goggles up and away from my eyes because they are cheap goggles and distort my vision. I cannot afford a good pair, and Logan will certainly not lend me his. He is a strange man in the matter of giving. At times he is a miser and at other times his very heart is yours.

Logan is my so-called employer, although paying me a regular salary would never enter his mind. He owns this beautiful Ryan Sport Trainer in which I now meet the dawn, and he owns two other airplanes which are my lot to fly according to demand and circumstance. One of the other airplanes is a cabin Waco. It is a biplane of sedate manners, but Logan rarely allows me to fly it because it is used principally to transport elopers to their rendezvous with fortune. Sometimes these are couples who act like newlyweds but are actually married to persons invisible. So it can be a delicate business requiring the utmost discretion. Technically, Logan trusts me to fly such people to Nevada or Mexico and return them safely, but he has told me why he prefers to pilot such flights himself.

"You brood," he recently explained. "It's bad for repeat business."

Logan never broods. He is a short, jolly, knob-nosed man with a consuming passion for tall women. He does not seem to care whether a woman is ugly, beautiful, or just middling, so long as she is long limbed and towers over him. I have wondered if there is perhaps something strangely pathological about his interest. Some of the women I have seen Logan pursue make me wonder if he sits on their laps and nurses at their breasts. Tall women and cigars are his only vices. He does not drink. And he would be a very good employer if he spent more of his money on engine maintenance instead of tall girls. His romantic preoccupations have made us both expert at forced landings.

Normally I spend my days playing shepherd to Logan's third airplane, which might easily pass for the world's ugliest flying device. It is called an Aeronca, and the forward end of its fuselage sags in a fat marsupial pouch so that the whole craft looks like a dejected pelican. It has a two-cylinder engine which would sound better on a motorcycle, and sometimes I am convinced the Aeronca would fly as efficiently if powered by a rubber band. I am rather ashamed of everything about it and never mention its existence in conversation. If a professional acquaintance asks me what I'm flying I name either the Ryan or the Waco. I pretend I have nothing to do with the Aeronca. It is only a few feet away but I do not even know it is there. It is like a half-wit relative.

Yet the Aeronca is my basic livelihood—such as it is.

In this sorry contraption I am supposed to take innocents aloft and show them what their planet looks like from an altitude of five hundred feet or less. They will not be taken higher because the Aeronca climbs very reluctantly with the weight of two persons, and the duration of the flight must not be over four minutes. That, Logan insists, is enough for two dollars. So it's once around the field and down. One of the dollars goes to Logan and I keep the other.

People who can be enticed into making such flights are known as "hoppers." On weekends if the weather is fine there may be as many as thirty hoppers a day. But during the week I become so lonely I sometimes find myself talking to the Aeronca. It helps to pass the time while I wait through the week for customers who either can't find the grass field because it is the same uninspiring parched color as the surrounding California geography, or they are too busy doing other things. I cherish solitude because life has so far been recklessly lavish with blessings for me, and I have yet to know the value of company in grief or agony. Yet the overwhelming loneliness of that little field sometimes threatens my reason. It is unproductive. The grass is tawny rather than green, always dry, very sparse, and stiff. There are large patches of sand and gravel everywhere, and so the slightest wind stirs up miniature dust storms and I must either stand up or go sit in the Aeronca, which is not very comfortable unless I trouble to hoist the tail and set it on a box. Then the seat is level and I can avoid a stiff neck while studying a book by a Russian named Jordanoff who has a way of describing instrument flying I find fascinating.

We call that little field the "south forty," although it is not nearly forty acres and it is only south of a railroad track and a complex of cattle corrals which seem to have been abandoned. We call it the "south forty" to distinguish it from the main airport where Logan holds court with his better airplanes and rangy concubines and where an outfit called Lockheed is building some real airplanes, obviously to be flown only by my betters.

When nearly a whole day of "south forty" has passed and I have not made a single ascension, I have caught myself taking up my pencil, holding it like a flute, and tootling at the Aeronca like a true shepherd. Or a snake charmer. For I have also caught myself asking the Aeronca when and how it plans to kill me.

"Go ahead, you miserable underpowered little sonofabitch! Try it!"

I am supposed to teach students to fly in the Aeronca. It would be difficult to find a more unsuitable aircraft for instruction, which is one reason we have so few students even at the bargain price of ten dollars per hour. Which includes my services. Six dollars for Logan—four for me. Instructor and student sit side by side in the Aeronca, so conversation is relatively easy, but there is only one set of controls. Hence I must either talk a student through a maneuver or reach across the cockpit and between his legs to make whatever corrections are necessary for survival. Sometimes the result becomes a grim wrestling match if the student is a complete neophyte, or if the engine quits for one reason or another and I must take over for a quick emergency landing.

The few other instructors I know have each made their own curriculum. I prefer to start my students with at least a spin and a loop on their first flight—thus discovering their "aptitude" for the business.

Some of our students have made one flight and never returned. I am beginning to wonder why and find it easy to blame the Aeronca. With two persons aboard it makes a terrible fuss about performing any but the most gentle maneuvers.

A very lucky week with the Aeronca will net me forty or fifty dollars. But the average is closer to thirty dollars, which I have told Logan is one reason I do so much brooding. Hunger, I have explained to him, pasties the face and glazes the eyes. Yet Logan suffers from a strange affliction. At the mere mention of money he suddenly becomes so tone deaf that he couldn't possibly pass a flight physical. His eyes take on that faraway look pilots are supposed to have, and I am convinced he not only can't hear me —he can't see me.

However, Logan is alert enough to know other ways to pacify his only employee. He is generous with promises.

It is true that even now he is making preparations for what he says will become the world's greatest air circus. I will be one of the pilots, and we will perform at county fairs and other possible occasions where the spectacular is valued. There are other circuses in America, and we know the pilots make as much as two or three hundred dollars for a single day's work. Or so they say.

Logan has flavored his promises with the tangible. Last night, perhaps touched by my forlorn appearance and the fact that I had not left the ground all day, he advised, "Your acrobatics* need smoothing. First thing tomorrow morning why don't you take the Ryan for an hour or so and wring it out?" This is the fifth time he has allowed me to play with his favorite toy.

There is one thing about Logan few men would dispute: He is a genius at acrobatics, in a class with Udet, Tex Rankin, and Paul Mantz. I treasure his slightest suggestions and try with only passing success to imitate his consummate skill. The Ryan is a wonderful partner, and while it is difficult to picture Logan as an extraordinarily sensitive man, his aerial ballets are more than thrilling. They are pure art.

The Ryan is a low-wing monoplane powered by a Menasco engine. The fuselage is of metal, the wings and empennage of fabric; it is capable of any acrobatic I know and forgiving enough to rescue me from certain cavortings I do not know, yet persist in trying. Loops in a Ryan are routine, of course. I can go around repeatedly making the circles round instead of egg shaped and easily keep the identical top and bottom altitudes for each revolution. The Ryan also spins smartly and without viciousness. Absolute accuracy in recovery is easy. Snap rolls can be repeated without loss of altitude again and again until you work off a mood or find it necessary to stuff your eyes back in your head. Symmetrically smooth slow rolls are more difficult. I cannot seem to manage one unless I start with a dive, which robs the action of grace. But with more practice and imitation of Logan I hope to learn.

The sky is now turning cobalt, deepening, inviting me. I have never been able to execute a decent Immelman, the maneuver named after its creator, a German fighter pilot of World War I. Somehow it is one of the most graceful and satisfying of aerial acrobatics. You begin with a slight dive as you would a straight loop, then climb at full power until you are on your back. But at the top of the circle, instead of cutting your engine to fall away and complete the loop, you hold full power and roll a half-turn to

---

* The word "aerobatics" now in common usage must have sounded too fancy for us. We never used it.

the right or left. Thus you emerge at the top of your original half-circle in level flight and flying in the opposite direction.

Logan is a master of the Immelman. I usually complete the maneuver technically, but my execution is without rhythm or true polish so that the effect is as jarring to me as it must be to any spectator. And occasionally I have stalled out before I can complete the final roll and have fluttered away in great upside-down embarrassment. It is like attempting an arabesque and falling on your ass.

I pull down my goggles because the instant the Ryan is on its back various bits of debris will certainly find my face. It doesn't seem to make any difference how many times I sweep the cockpit flooring; once the Ryan is on its back, always, from somewhere, pebbles, dust, gum wrappers, or even one of Logan's cigar butts will find its way through the various openings and crevices and float past my face. I have learned to hold my breath during the first roll and keep my mouth closed. It's like inverting a wastebasket. There's always something left in the bottom.

Now full throttle and upward while the cobalt sky slides past the propeller blur and disappears behind the wing. Blood is coursing to my head, and my eyes wait for earth's rising horizon.

There! A thin chocolate line slicing through a scarlet sun. The sun is hanging down from the horizon and I am hanging against my belt and my feet tend to fall away from the rudder pedals. The engine begins its sputtering, which it always does when inverted. Now, not too abruptly, an easy roll to the left while holding the nose on that glowing target of fire. Speed? You have enough.

Roll for a count of four. Completed. Yet I am unsatisfied because there was a hint of trembling which meant I used too much rudder and skewered instead of rolling. Still it was better done than usual, and I throttle back to analyze and prepare for another.

When the Menasco's fuel system is again stabilized, I roll the Ryan over on her back, pull the stick momentarily, and "Split S" into a vertical dive to six thousand feet. The beautiful wilderness aloft is now forgotten. The Ryan and I are open for the morning's business. On the way down, while the Menasco resumes its sputtering, I consider dutifully the folly of acrobatics without parachute. But Logan does not have any parachutes and I certainly cannot afford one. And Immelman, who was being shot at while he performed similar maneuvers in a much less stalwart craft, certainly did not have a parachute. Of course, Immelman was killed.

All right. *That* has been thought about.

There follows a half-hour of determined practice during which the sun's new flaming is ignored and the sky merely serves as the reverse plate of the ground haze. And most of the dust that is going to be shaken out of the Ryan this day has found its way to my face and ears and jacket. My fingertips are numb because in my anxiety for perfection I have tensed too hard on the control stick.

Before my time they ordinarily called it the "joy-stick." I wonder again if the label was simply the result of phallic suggestion or inspired by comparison with the only comparable physical endeavor. It is astonishing how much lustful joy can be found aloft.

Logan will be waiting. It is his fuel I am consuming. So now I must go down to him, delaying the end in a series of graceful chandelles and Cuban Eights.

Goggles up again. I breathe deeply of this enormous morning. Now the mountains lie in clear, rumpled piles against the lower sky, and beyond them gashes of yellow white mark the desert. I do not care about my ever-present economic problems. It is a morning for joy. There are eagles soaring about those mountains who, for this moment, dare not compete with me.

At one thousand feet I submerge into another environment. The sparkling clarity is smothered in haze and there are smells. There is movement in the streets below as I approach the airport and circle to land. People are going to work. At once I am again one of those people and my exaltation vanishes. It is one minute before eight o'clock of a Friday morning.

I slide through the murk passing over my miserable back-lot "south forty." I continue across it and land moments later at the regular airport. Logan is waiting for me, his smile genuine and warming as always, his somewhat defiant stance tempered by his curiously wistful manner. It must be that tenuous, ever so mischievous little-boy appeal which initially endears him to his legion of Amazons. I know him to be a little bantam cock, vastly confident, although without conceit. And I know that although his innate cunning is much given to schemes against the chastity of

American women in general, he is no mere libertine. No endeavor—physical or mental—can remove the honest courage from Logan's eyes or mar the special beauty of a face punished by a thousand winds.

"I was watching you. You're improving."

Thank you, almighty mentor, for sweet words in the morning. Feasting upon them, I forget all of your failings and would rather fly for you than any man on earth.

"Sunday," he continues, "you will be in charge here. I am taking the Ryan to Oxnard. There's an air show."

"What about our circus?"

"That will come. I'll make some contacts in Oxnard."

He will indeed. All tall. Meanwhile, who am I to complain? He will be leaving me at an honest-to-god airport with two modern airplanes to lure the customers. And on a Sunday. I cannot help but prosper.

"Now get on over to the 'south forty' with the Aeronca and hang this new banner along the fence."

He unrolls a long piece of canvas. The legend upon it is in large letters and simple enough: AIRPLANE RIDES $2.00. The words are in black and the numerals in red.

"How does it strike you?"

"I think it's very pretty."

"The guy who made it is coming over for a free ride this morning. Give him ten minutes."

"Okay, if you say so, but I don't think the sign will do much good."

"Why not?"

"The grasshoppers can't read and over there no one else is around."

Logan grunts. He has not really heard me, yet, again, I must try my pitch.

"Why don't you let me work out of here? I could do five times the business. Good fishermen go where the fish are."

Again Logan must play the abused general. I predict his refusal and it comes on cue.

"No."

I know very well why his refusal is so automatic. If we are both flying customers out of the same airport, we are not expanding; in Logan's peculiar business logic, he would be competing with himself. Furthermore, he would have to pay additional airport charges for whatever space the Aeronca and I might occupy plus a certain percentage of the monthly gross. Over on the "south forty" there is no charge for anything. It is just there.

I roll up the banner and toss it in the Aeronca's open cockpit beside the paper bag containing my lunch. I cast off the tie-down lines and climb into the cockpit. Logan ticks the prop around three blades and waits. We both know the Aeronca's engine is a crosspatch in the early morning and is both difficult and dangerous to rouse unless the positioning of the planets and the tilt of Logan's cigar are exactly right.

Logan yells, "Contact!" and I turn on the magneto switch. He yanks down on the little paddle blade, and my humble ship vibrates with life.

"Better bring the sign back tonight. Somebody might swipe it."

I nod. In three minutes I arrive at my office and the waiting begins.

I wait all morning for the banner painter. But there is no intrusion upon my exile. Perhaps he cannot find the "south forty" because he has taken the road along the power lines which constitute such a remarkable flying hazard on the north and south that I fly under them on either approach or climb after takeoff. If so, he could easily miss such a low and cringing plane as the Aeronca. Or me—in the shade of its wings, pining for action. His own banner would be no help. I have dutifully secured it to the fence along the road, but both are so far away and there is such an expanse of mesquite in between that only a previous customer would be likely to spot the Aeronca's wing.

One of these is Nelson, a tall, extraordinarily easygoing man who can barely fit his bulk into the Aeronca. He looms on the desolate horizon in mid-afternoon, and I watch a vague heat mirage behind him as he proceeds toward me.

"Pretty warm for a Friday."

"I should better be playing golf."

We squeeze ourselves into the Aeronca and he sighs.

"My sensible friends think I'm crazy doing this."

It is Nelson's second lesson and he flies very well. I tell him he has the qualities of a professional, which seems to please him. When I also tell him of my income he is amused, which in turn pleases me because I do not want him to think I seek sympathy. I anxiously reassure him.

"Sure. I might make more money at something else. . . ." I wave my hand across the baking valley as if it all belonged to me. "But what is better than this?"

At this temperature and load, the Aeronca is barely capable of flight. It takes me several minutes to recover from the miracle of our takeoff, which would never have occurred without the ten-mile southerly breeze.

The majority of the lesson is devoted to keeping us from falling out of the sky, and I debate the wisdom of advising Nelson he might go elsewhere and fly a real airplane.

Linsk is waiting for us on the ground. Nelson has persuaded him to take a lesson. The direct physical opposite to Nelson, he is slight and wiry of build, delicate handed, and nervous. He regards the Aeronca with deep suspicion.

"Is this the best-looking airplane you can offer?"

"Beauty is only skin deep."

"Is it safe?" He is trying to make light of his concern, but his doubts are obviously very real. "It only has one wing."

"No, two. One on each side. You might as well be in your mother's arms."

"*My* mother did not raise her boy to become an aviator." He crawls reluctantly into the cockpit. "Just remember, I am afraid of everything. And heights make me throw up."

Once again the Aeronca manages to become airborne—climbing at a livelier pace now because of Linsk's lesser bulk. And I proceed with my indoctrination, admiring his grim determination to master something so totally foreign to his nature. When at last we land his face is white and damp with perspiration, but he has not thrown up. I congratulate him.

"Thanks. I learned a lot."

Did you really, brave man, between those bouts of abject terror?

"I'll come again tomorrow if your nerves can bear with me."

My nerves? Bless him. I watch his trembling fingers as he lights a cigarette, knowing already that he will never conquer flying or his fears. Now he will return to his own world and tell his friends of his wonderful discovery. And he will verbally work himself into a venturesome box from which there is no escape without ridicule. Perhaps if I urge him to persist he may even one day kill himself. And yet—?

"Sure. See you tomorrow."

Sunday morning. I flip the Ryan's wooden propeller four times clockwise then glance at Logan, whose head projects just above the cockpit cowling. He bites on the stub of his cigar and nods.

"Contact!"

I reach for the blade with both hands, swing my left leg across my right, and heave downward.

The Menasco engine catches instantly and commences its characteristic flat bleating through four short exhaust stacks. I move carefully around and behind the propeller to pull out the wheel chocks.

Logan sits quietly, warming the engine, and once he looks out to wink at me broadly. I suppose he means that this day he is going to put on such a show at Oxnard the future of our air circus will be assured. I can see only great good in the form of fame and fortune resulting from such an endeavor. Logan and I will rise above the ordinary.

I wait near the tip of the wing while he fastens the chin strap of his leather helmet, tightly as he always does so that his ruddy cheeks become muffins of flesh. Once again I become aware of a curious and great affection for Logan which makes of him much more than a parsimonious employer. We are, after all, comrades-in-arms, trying to make a decent living in a very special way. Studying him now, clad in his leather armor, all business as evidenced by the firm clamp his lips have formed around the cigar, I can forget his failings. I see him as a magnificent craftsman-artist, carving his intricate designs against the sky.

When Logan taxis away still chewing on his cigar, I wait until only the back of his helmet is visible. For I do not want to embarrass him or have him think me a young and sentimental fool hopelessly in love with all things and people related to flying. Logan probably would not understand why I am suddenly and irresistibly compelled to flip him a little salute.

Business is good all through this Sunday. I fly the cabin Waco almost continuously—up, around, and down. "Thank you, sir. The Pacific Ocean? That way. The Douglas plant is beyond those hills. We're too low to see it. The movie studios? Just off the right wing. There is Warner Brothers. Thank you, ma'am. Come again."

The only time I have for lunch is when the Waco needs refueling. While the truck takes its time coming from the other end of the airport I chew on a sandwich and fret. Some of the potential customers who have been in an almost solid line along the fence are drifting away.

I am busy until late afternoon and then, as always, there is a sudden inexplicable desertion from the fence. People's Sundays are regimented, and the roast is in the oven. Yet I am more than satisfied. My share of the day's work is fifty-five dollars. Glorious Sunday! About now Logan should appear smiling for his share.

I taxi the Waco to the hangar and put it to bed. Then I wait in the dusk for Logan. The sky becomes shaded with a lingering emerald, and very high there is a long procession of miniature cirrus cut like the headgear of Sistine nuns. Come on back to the stable, Logan.

Logan has never equipped the Ryan with landing lights. Too expensive. "Airplanes are worse than broads. You fancy 'em up and maybe ruin 'em."

There is the last of the light. And in minutes it will be night. I suppose a good night for you, dear philandering boss. Something much more important than money, say a redhead no less than six feet tall, detains you in Oxnard. Of all places.

Never mind. I will keep your money until morning.

I will keep your share forever, old comrade. Because while I am standing waiting in the twilight you have been dead for more than three hours. You gave them a show all right—finessing with one of those long, perfectly controlled spins which before you have always held deliberately until those who understood a spin began to squirm uncomfortably and those who did not understand a spin opened their mouths and gasped. And this one time (there is no second time for such things, of course), you must have hesitated three or four seconds. Or maybe more. I do not know because I was not there. I was in charge here, boss.

Something went wrong during those last seconds, and the Ryan hit the earth with a terrible muffled clump and the explosion of dust was awful. That is what they tell me. Your original spin must have developed into a flat spin because the Ryan hit that way and folded around you like a crumpled flower. Instantly, they say. And no fire.

Well, that's something. May they all be ten feet tall and beautiful, friend. Tell me, should I sit down and bawl, which is what I yearn to do, or just shrug my shoulders helplessly and say, "Well, hell. . . . that's the way it goes . . . ?"

Well, hell? Wouldn't you say that? If you were still young and inexperienced at this sort of thing and hence quite uncertain about the proper form? You neglected my education, boss. These things take practice, and you, Logan, happen to be the very first friend I have lost to the sky.

And Jesus Christ almighty, how suddenly I hate it!

Chapter Two

Notice to Airmen

*T*here are those who consider these supersonic times the halcyon days of flying. They are and they are not. There have been other days, much simpler, more immoderate and infinitely more joyful for those whose very life and love was flight.

Two generations were changing places when the old hard-bitten barnstormer was commencing his rather pathetic decline into almost total obscurity. Even though many had trained to fight the first war in the air, few people honored them and almost no one in aviation realized the relative scarcity of their numbers. In America, for one successful Rickenbacker or Reed Chambers there were hundreds of ex–World War I pilots still trying to scratch a living from the skies in any way they could. The same applied to Germany, which had emerged from chaos into the arms of Hitler. The Richthofens were long dead, and only Udet survived to remind Germans of aerial glory. There was an official flying school for the very select sponsored by the new Lufthansa and, incidentally, the Luftwaffe. The vast majority of other flying was done in gliders by sport groups who spent an inordinate amount of time at calisthenics and marching.

For several reasons the survivors of the Royal Flying Corps fared better in their native English skies. At the end of the hostilities they were in relatively greater numbers and their educational level at original selection was higher than in the United States and most other countries. British pilots seemed to find it easier to keep pace with the rapidly growing state of the art.

France, with almost no reservoir of privately inspired pilots, depended almost entirely upon her military veterans, and the whole French aviation endeavor between the wars bore a quasi-military look.

Perhaps the greatest factors influencing the training of flight personnel everywhere were the relative capriciousness of power plants and the still primitive ceremonies involved in forecasting and distributing weather information. In the United States, by the time a student soloed, he had not only been given repeated practice in forced landings, but his chances of experiencing a few genuine emergencies were very real.

Anyone who had achieved a transport license could teach flying, although in government-approved schools an instructor's rating was also necessary. Most instructors had their own very firm idea about the right way to fly and what a student should know before being sent off to join the birds. The majority of instructors secretly considered themselves the world's greatest aviators, nothing less and nothing more. They only deigned to teach, and since they had little truck with the psychological effect of their dicta, their students endured a hard apprenticeship.

Nearly every instructor was hardheaded about his beliefs and was likely to salt the air with picturesque phrasing when suggestions were made that his discipline was bad for repeat business. The concept that anyone could be taught to fly was simply not in them.

Even by current standards the price of learning was still very high. During the year 1931, when the dollar was sound and round, the average charge at approved schools was:

Private pilot	$550.00
Limited commercial pilot	$1,207.00
Transport pilot	$4,255.00

In contrast, room *and* board at such schools averaged twelve dollars a week.

Less affluent students, determined to learn flying the hard way, could buy lessons at almost half the approved-school rates by trading their labor —sweeping hangars and polishing airplanes—for the local fixed-base operator.

Until the latter-day thirties aeronautical innocence and ignorance abounded. Flying was roughly divided into four categories: the military, which was almost entirely devoted to "Pursuit" and "Bombardment" (there was no formalized transport arm); the airlines, which all together employed less than one thousand pilots; the barnstormers, which included skywriters, exploration fliers, record-setting personalities, fixed-base operators, movie and news pilots, and whoever else tried to persuade themselves they were harvesting a decent living from the sky; and, finally, the occasional private pilots whose numbers about matched the roster of Americans playing polo and who needed similar funding to pursue their sport.

There was no Air Traffic Control as such until the late thirties, and even then it consisted of a crude arrangement employed almost solely by the airlines and often ignored by them in good weather.

Everyone else flew as they damned pleased. Official "Contact" flight restrictions were approximately the same as for visual flight rules today, although few pilots paid the slightest attention to them. If the ceiling over New York, Boston, Minneapolis, Chicago, Los Angeles, or Memphis happened to be 200 feet, you flew across the city at 100 feet and only very rarely was a violation filed. No self-respecting second lieutenant in the Air Corps would miss a chance to "beat up" his girl friend's house with a proper buzz job, and even some airline pilots treated their few passengers to low-level passes so the wife (or girl friend) would know their man would soon be available.

In America, ignorance of true instrument flight, except for the airlines, was almost universal. Only a few schools, notably the Boeing School of Aeronautics, Spartan, Parks, and the several Curtiss Flying Services made any serious attempt to teach instrument flying. Link and his revolutionary ground-fixed trainer were still unknown. The average instructor ignored the turn and bank instrument which adorned the panels of the later aircraft, if only because the seat of his pants told him when a turn was not coordinated and because any man who spent more than a minute or two in a solid overcast was asking for trouble. If a pilot was stupid enough to be caught on top of a solid overcast then he had better have his wits about him and a good watch. Once trapped he was advised to hold a steady compass course for his destination (or the lowest known terrain), and when the proper time had passed, commend himself to God. Then he should start a 45-degree dive and hold both compass course and airspeed until hopefully he broke out underneath. Pilots who survived such letdowns soon developed a mastery of the 180-degree turn.

The average airman's mechanical skills and overall knowledge of aircraft functional elements were perhaps better than that of the present generation. A great many pilots maintained their own aircraft. Mechanics who devoted themselves totally to maintenance were often near geniuses of improvisation, for the supply line of parts and the availability of money to pay for them were both extremely tenuous. Even so, many "mysterious" forced landings and crackups resulted from ignorance of such elementary factors as carburetor icing.

Electronics were simply nonexistent in the average airplane.

In England and Europe licensing was already a long and complicated process with most applicants previously selected as likely to pursue professional careers. In America the licensing requirements were not so comprehensive, but hardly easy. Even aspirants for a private license were required to demonstrate sideslips, engine-off landings, and recovery from spins in addition to ordinary precision maneuvers. The emphasis was on practicality rather than written questions and answers. Transport licenses required the applicant to demonstrate (among other proficiencies) *very* precise recoveries from two turn spins. Usually the inspector, who was actually employed by the CAA and not a designate, considered it his duty to go aloft with the applicant, and the slightest deviation from perfection brought a sharp rebuke. Any sort of clumsiness, mental or physical, meant failure.

Everywhere airports were few and actual runways hardly existed. Control towers did not appear at even the major terminals until the mid-thirties, and the occupants spent more time pointing their green lights at aircraft than on the radio. It was up to the pilot to select his landing direction according to the windsock or perhaps one of the new wind-tees. Take-offs were made when and how any pilot desired, except at major terminals where the more conscientious pilots *might* wait for a green light from the tower.

Navigation was almost entirely *avigation*. The airlines knew about the new radio beams and had the equipment to use them. Very few others even knew of their existence. The government charts were good although a nuisance to manipulate in an open cockpit, and there were as many different and special ways, as there were pilots, to fold and paste charts to avoid confusion. There were published route directions which might read as follows:

PITTSBURGH-DAYTON

Magnetic course	Dist.	
249½	47	Pittsburgh (Mun☼) (▪) From ☼ on W side of R opposite Mckeesport; 7 mi cross HW & RR; 9 mi Canonsburg; 8 mi cross HW 3 mi NW of Washington; 23 mi to Wheeling (06 mi N). a/c W.
W	115	Wheeling (▪) Follow HW for 46 mi; pass 1 mi N of Cambridge☼; 21 mi cross R & RR; 7 mi cross R & RR; 22 mi cross R & RR diagonally; 19 mi Columbus Mun ☼ a/c 265½°
265½	71	Columbus (▪) 6 mi into Columbus ‖ HW on N side for 43 mi to Springfield (☼4½ mi SE) follow HW 22 mi Dayton☼

Legend:	☼	Lighted airport	▪	lighted airway
	O	Airport or landing field	R	River
			RR	Railroad
	‖	Parallel		
	HW	Highway		
	a/c	Alter course		

Of necessity airline pilots were particularly ingenious in storing necessary information about their routes so it could be used while in flight. One of them, Elrey Jeppesen, a United pilot, kept a methodical notebook which eventually became a worldwide institution of inestimable value to airmen of every nationality.

All of these things were true only a few years ago. Some of us who have survived through the three concurrent generations and are still flying are sometimes appalled at the swiftness of aeronautical change. Most of it has been for the good if you are willing to accept crowded skies and the ever-lessening romance for general public participation.

Passenger or pilot, better off or not, flying is no longer a gypsylike, adventure-promising way of life, and never will be again. Now, through the magic of Robert Parks's brush, surrounded by a few words, perhaps we may recapture at least some essence of aeronautically less sophisticated times.

Chapter Three

The Silver Wing Service

Paris in the spring! A May morning of the year 1927. Imperial Airway's newly launched "Silver Wing Service" gives eighteen passengers who boarded in London a dividend on their £9 (U.S., $44) ticket—a bird's-eye view of the Eiffel Tower before a turn is made back to the northeast for a landing at Le Bourget.

The aircraft is an Armstrong-Whitworth Argosy, of which seven were built. All were named after cities. This one, the *City of Birmingham*, has been 2½ hours en route from London; headwinds of a velocity of 30 kilometers have caused her to be behind schedule.

Captain O. P. Jones is in command. He sits on the left side of the open cockpit and has been in radio communication with Le Bourget, which has reported a ceiling of 100 meters (300 feet) and visibility of 1 kilometer (about ½ mile). No matter. Captain Jones is pleased with the behavior of his wireless, which this morning allowed him to converse with Le Bourget from a distance of more than 50 miles. And as for the landing he will execute an L-K (local knowledge) instrument approach. He will accomplish this one as handily as he has done so many times before. The procedure is as uncomplicated as his aircraft. He will let down to the top of the overcast after searching out a thin spot or a small hole with something he recognizes at the bottom. Almost anything in the general vicinity of Le Bourget will do —he knows by heart the compass course to the aerodrome from a church, a small factory, a certain road crossing, a rock quarry. And the land all around the aerodrome is conveniently flat and free of obstructions.

Once on course he will throttle back the Argosy's three 385-horsepower Armstrong Siddely Jaguar engines until his airspeed drops to 60. He will keep wings level with the aid of a new-style artificial horizon and keep a straight course down through the muck with his turn indicator. Warning of an incipient slip or crab will be signaled to him instantly from the feel and sound of the damp wind on his cheek—a sensitive instrument which long ago won his faith. Since his war service with the RFC he has become quite proud of his cheek and, like his comrades on Imperial Airways, is less enthusiastic about future aircraft designs which show the pilot enclosed in a glass cockpit.

"Bloody chicken coop!"

If Captain Jones should miss his descent-distance plan slightly, the error is not serious. There are no runways at Le Bourget to trouble a man; like Croydon, you can land anywhere on the aerodrome and on a still wind morning like this, in any direction. White lines are marked on the green grass, and even in wretched visibility they inform a man whether he is landing too short or right on, or suggest it would be wise to circle for another pass. No flaps, no spoilers, no retracted undercart to fret about. Just chop the throttles, ease the nose up, and settle gently to the greensward at a nice 50 miles per hour. A final hissing as the wheels brush the grass, then a muffled thump. "Are we down?"

When there is no wind what could be sweeter than a wing loading of 5 pounds?

And the passengers are delighted with the *City of Birmingham*. While their American cousins who have managed to fly from one city to another in a small single-engined biplane are lucky to have a mail sack to sit on and a ham sandwich to chew while they mentally review their wills, these passengers have enjoyed a modest but tasty lunch. It is enhanced by a fine wine and served with the practiced skill of a steward who has learned his business caring for passengers aboard a blue-ribbon Cunarder. As a consequence they are so full of zest and the joy of accomplishment they dare to attempt conversation over the high decibel count in the cabin. England, once lord of the seas, now promises to rule the skies, and the competition is weak. There is KLM with their Fokkers (very Spartan travel compared to an Argosy), France has her Air Union flying LE 021s, and the Germans are still in the Great War's shadow with their single-engined Komet IIIs.

As the Argosy passengers survey the earth below and the curtained, candelabraed cabin which is their world aloft, and as they study the large airspeed indicator and altimeter placed in the cabin for their interest, they have reason to be smug about Britannia's place in the aerial scheme of things.

Soon they will have more inspiration, for by 1929 Argosys will be flying through to Egypt with such distinguished passengers as Sir Samuel Hoare, and Edward, Prince of Wales, shepherded by the same Captain O. P. Jones. By November 1929 Argosys will be showing the flag all the way through to Karachi, and the Air Ministry will have signed an agreement with Imperial Airways calling for a subsidy of £335,000 for the European and India routes. During the same year GAPAN (the Guild of Air Pilots and Air Navigators) will be founded, "to further the efficiency of commercial aviation and to uphold the dignity and prestige of air pilots and navigators."

For the British the promise of the skies is on every horizon. They are as temperamentally suited to it as they ever were for the sea. In the Royal Aero Club and in the furthest Dominions where proud British flying men gather, there is no doubt that the sun will ever set upon their flying empire.

Chapter Four

New Year's Eve

*F*lying the U.S. Mails in an open-cockpit De Havilland-4 is a job for young men already matured beyond their numbered years. For they know what it is to be but one individual aloft, utterly solitary except for their secret fears and carefully monitored hopes. On this last night of the year 1926, only the moon will serve as one young man's companion in celebration. Yet he has but the one all-consuming complaint of his lot, a deep elemental yearning for heat to convince him that he is really twenty-seven years old and not some long-embalmed mummy with the warmth of life forever gone from his bones. Once more he has discovered a basic fact of aerial life: *There is no place in all God's creation colder than an open cockpit at midnight.*

Below in the moonlit void pinpricks of light wink at him all along the route from Omaha to Cheyenne. They are acetylene navigation beacons set out by the government to guide this post office employee on his way. Below in the depths the great western plains of America are still sparsely settled, and in the wilder regions there are still as many Indians as white men. There are few fences and only a few nearly deserted two-lane roadways.

NEW YEAR'S EVE
De Havilland DH-4 at night over the North Platte River, Nebraska

Where the land has been cultivated there now remains only winter stubble to relieve the barren earth.

Soon this young man will come upon the lights of North Platte, and then far beyond the horizon will be Cheyenne. And beyond that frontier and yet many horizons further, some of his cargo of mail will be flown through to Salt Lake via Rawlins and Rock Springs; some of it will be carried on to Marina (Crissy) Field in San Francisco, and some transferred to Western Air Express where a pilot named Jimmie James may fly it over the mountains to Los Angeles.

It is a good life except for the cold. A man can make as much as six or seven hundred dollars a month if he is lucky. It is a good year if you ignore the 155 forced landings the airmail service experienced because of mechanical failures, or the more than seven hundred because of weather. There have been two pilots killed, and fifty-nine have suffered minor injuries. Nine planes have been lost and well over two million miles flown. It has been a good year—for the survivors.

Among those are the elite of the flying world. Who would not be proud to name as his comrades such men as R. T. Freng, Sloniger, the quiet-spoken Ed Matucha, Shirly Short, the easy-laughing Dinty Moore, Walt Adams, towering Dean Smith, Jack Knight, and the redoubtable E. Hamilton Lee? Some have as many as four thousand flying hours . . . and some admittedly, an honest four hundred. But the average is about 2,500, which is a lot in these times, and if a man keeps his flying wits he is certain to acquire more.

Yet there are rumors all along the line. It is said 1927 will be the last year of government operation. The Kelly Act calls for the task to be performed by private companies; and with a supposed 80 percent of the revenue going to the contractor, big money is taking an interest in matters aeronautical. The rumors claim they will certainly want to stuff some passengers in with the mail.

Well, not in this airplane, which is strictly war surplus. It was designed as a bomber and/or observation craft. This one, like the night of this year, is old and rather weary. It has been more repaired than modified, has its original 12-cylinder Liberty engine, and still wears a wooden prop instead of one of the new Curtiss-Reid all-metal spinners. The tires have not been replaced by the new over-sized variety, and the exhaust stack has not been carried all the way back past the cockpit. Never mind. A man can judge his engine-fuel mixture by the color of the flame and without getting a stiff neck trying to watch it.

The unique aroma of the exhaust mixed with the chill night air is somehow comforting. It matches the reassuring thunder of a Liberty engine at work. Back in Omaha it had taken three men hand in hand crack-the-whip style to start it. After they had caught their breath enough to curse, they said they hoped to God the improved Libertys with drilled pistons and an improved oil pump would be easier to start. And maybe the plumbing would be less exasperating.

A flying man has been faced with all kinds of changes this year. There are radio stations at every terminal field to pass on the weather. The boundaries of emergency fields are outlined by small white lights only two hundred feet apart. The terminals will turn on a better than three-million-candlepower floodlight to make your night landings almost too easy, and even if you have to set down out here in the wilderness you have parachute flares to light up half a section. And right now they are replacing the 18-inch rotating beacons which are every fifteen miles along the route with big 24-inchers. How could a man become even half-lost? Easy. Just bring on a low overcast.

It's been a good year. The DH-4 has carried you and 500 pounds of mail through a lot of muck aloft and earthly tribulation below. She can be a bitch in a crosswind and calls for some muscle in rough air, but once you have her trimmed she becomes a very stable bird. Of course, she is no Douglas M-1, which was built specifically for the mail service and went to work just this year, but a DH-4 is far and away a better flying machine than a "Jennie" or a Standard for the same job.

On a night like this it almost seems a man could fly on forever—if it were not so cold.

Chapter Five

Old Number One

*I*t has been a few years now since Slonnie throttled back and sideslipped away forever. Of course it was time if we are willing to accept the biblical increment of three score and ten, but those of us who laughed with him and sorrowed with him and flew with him still feel cheated. For Slonnie was not only a true airman, but also a legend, and in time he became an institution. His real name was Sloniger and he is, or was, *the* airman I profoundly wish more pilots and people who find exhilaration in just looking at the skies could have known.

Obviously I am having trouble referring to Slonnie in the past tense; perhaps because his mandates are still so very much alive. In a way he was a sort of Moses who left his aeronautical precepts carved forever on our thoughts. The separate employment of the terms "airman" and "pilot" is in itself a categorical heritage left by Slonnie. There is and always has been a considerable difference, and those who had the fortune to fly with Slonnie knew their chances of qualifying as airmen would be mightily enhanced.

We are now in our third generation of pilots, a vocation or avocation pursued by so many human beings of every conceivable description that the multitude is often more depressing than inspiring. These individuals are presumably capable of controlling a machine skillfully enough to cause it to become airborne, then proceed to a previously intended destination and land without serious damage to machine or occupants. In the United States a 2¼-inch-by-5⅜-inch piece of government issue paper which pilots are supposed to carry on their person *infers* but does not guarantee they can accomplish such an endeavor. It is officially called an "Airman's certificate," which is all too frequently a semantical rape of at least one innocent noun.

There are airmen and there are pilots: the first being part bird whose view from aloft is normal and comfortable, a creature whose brain and muscles frequently originate movements which suggest flight; and then there are pilots who regardless of their airborne time remain earth-loving bipeds forever. When these latter unfortunates, because of one urge or another, actually make an ascension, they neither anticipate nor relish the event and they *drive* their machines with the same graceless labor they inflict upon the family vehicle. Slonnie was not very tolerant of such people.

Slonnie preferred the company of airmen, and since his seniority number with American Airlines was *one* (which was certainly all the regal juice a man could have), he was able to choose his peasants. Any man flying for that airline was presumed to know him, which did not necessarily mean he should know you. My own seniority was considerably more humble than number one, and it so happened that after four years with the line I knew Slonnie by repute although I had never actually met the great man.

Came World War II. Then one blizzarding night in Newfoundland, in the Pleistocene era of trans-ocean flight, I was at last presented to Slonnie. We were in the operations office checking weather and load when I was introduced to the noble of nobles, and I was instantly more shocked than impressed. For I knew he was about to perform the same duty with which I had also been impressed, which was to fly an aircraft across the winter Atlantic Ocean, and I was at once convinced this many-fabled hero had been drinking. Heavily.

I was better prepared for his physical appearance, which was sometimes compared to that of a Sioux Indian who might have put aside his war paint long enough to drop in on the local trading post with the intention of stealing all he could carry. At the moment of our handshake I thought any prudent factor would have immediately checked his whiskey supply. I tried to find some comfort in reminding myself that here was the first drunken Indian I had ever met who was about to fly the Atlantic on a dark and tempestuous night, and how many other men would ever stand witness to such an extraordinary event? At least, I thought, the renowned Slonnie's genial smile denied he might try for my scalp.

As we exchanged the usual empty phrases of strangers brought together and began to probe gingerly at the hidden man beneath the uniform, I noticed how Slonnie shifted his weight from one leg to the other in a sort of swaying motion as if he were instinctively responding to some distant tympani or, as I suspected, he might be trying to keep the operations office from precessing overmuch. I found myself trying to catch a whiff of his breath and was instantly ashamed of myself. Had I suddenly become my hero's self-appointed guardian, my thinking twisted by awe to acquire a little old lady's presumption of evil?

As if to confirm my scepticism, Slonnie now mixed gesture and word into the linguistic style used by all movie Indians when they are telling the big white chief how they shot down an eagle with only one arrow. What he was really saying was how remarkable it was that two Lincoln, Nebraska, boys would finally meet on this remote stone in the subarctic, but he kept grinning and slurring his words while his rather delicate hands performed chandelles and whipstalls and Cuban eights until all I could think of was how to find out which route he would fly across the ocean, whether it be via Iceland, the Azores, or direct, and make it my business to select a different one. There was then no such impediment as air traffic control to inhibit free style over any of the seas, and I wanted my aircraft to be flying several hundred miles out of his way.

If you will gradually exchange a snow-swept, ice-riven night for a gray Scottish morning you will find us both arrived quite sedately at Prestwick. We are in the RAF mess trying to down a wartime breakfast of bubble-and-squeak.

Slonnie has brought his crew to dine opposite mine at a long table, and we are all weary and grubby and given to faraway staring, having not yet truly descended from our long night of flight. Still spellbound in my personal heights I sip the lukewarm English tea and watch the rills and canyons in Slonnie's hammered bronze face. And I suppose, without caring now, that our Sioux must have been at the firewater again because his phrases are more than ever slurred and his hands continue their aerial circus as if the competition of a great ocean passing beneath them during the night had not even momentarily halted their performance.

Yet now there seems to be a more recognizable undercurrent in the flow of sound and movement. Now the easygoing, languid phrases and constantly moving hands have become messengers offering me intelligence which I can interpret; it is like suddenly discovering a new form of music and through repetition appreciating it more than you had hitherto believed possible. And in this curious auditorium with its drowsy audience the music reflects only the constantly recurring theme of flight, and so I am soon lost to its beauty. When I dare to speak I find myself anxious that my echo will be in the same key, for now my half-slumbering mind has advised me I am in the presence of a great airman who has forgotten more about flying than perhaps I shall ever know. And I realize that no, old number *one* has not been drinking; indeed I realize he would have scorned a man who might in any way dull his senses when even the prospect of flight was about.

Later, when we began to truly communicate, I found Slonnie had nearly all his lifetime been dedicated to those same precepts which I had long held dear. His coat of arms was trenchant—World War I flying rampant on a field of barnstorming, and crossed with a bar of China service when men with wings in that land were considered to have descended directly from the everlasting celestial beyond. Moreover, he had flown the American mails in open cockpit DH-4s, flying *mano-o-mano* with a certain Charles Lindbergh who was and *is* himself very much of a genuine airman. Both were employed on Robertson Airlines, which was eventually absorbed by American Airways, which in turn became American Airlines, and thus it was that Lindbergh went on to international glory while Slonnie became number *one*. They were both airmen hatched from a very special nest, although no one could ever possibly mistake Lindbergh for an Indian. Yet their easygoing hand gestures are much alike and Lindbergh, apparently still compelled by an identical metronome, takes on a similar swaying motion when he praises Slonnie's aerobatic ability.

Long after our gray-green breakfast in Scotland, fortune declared Slonnie should become my chief pilot, and as a consequence I was able to share a flight deck with him occasionally. It was while observing the master at work that I became convinced that no pilot could hope for airman status unless there was a strong sense of rhythm in his psyche. It made no difference if the aircraft had two engines or four—Slonnie flew them all as if they were an extension of himself. I was amazed and not a little miffed to witness the same great bull-hunk of iron I had been pushing around the sky apparently come to life and in Slonnie's hands behave like a buoyant creation of wood and fabric.

I never saw Slonnie make an uncertain approach, if only because he thought far ahead of his aircraft so consistently and was thus always in position to *start* his approach properly. It followed effortlessly and naturally then that his power discipline was never rushed and his descents timed in perfect harmony with terrain and wind. He somehow managed such smooth reductions right down to runway threshold that those in the cockpit had to check the gauges to reassure themselves all was in order, and until the moment of touchdown many passengers believed they were still cruising.

Slonnie would set his altimeter according to the pressure given by the local tower, but if we were in VFR conditions I never saw him so much as glance at it again. He *saw* his altitude rather than driving around the pattern on a set of numbers. His turns, either climbing, descending, or level, were a fluid result of airman and aircraft. They were positively oily during entry and recovery, and throughout, the altimeter hands appeared to be glued in place. Regardless of bank or degrees of turn, needle and ball always remained precisely right, and yet Slonnie kept his eyes out of the cockpit where they belonged. He sensed his aircraft as if it were a living entity needing nothing but the educated seat of his pants to whisper warning that a skid might be threatening or to hint of a puny drift in altitude.

Slonnie not only had rhythm but the most soothing gentility in handling both aircraft and people. His ever-moving hands flew for him all his waking hours, for flight was the very breath of life to him. Once he confessed rather shyly that he had not spent thirty consecutive days on the ground since 1917, and when he was my chief pilot he had already accumulated 25,000 hours the hard way.

Impressive as such figures might be, flying by the numbers was anathema to Slonnie. He would often wave a hand reproachfully across whatever airspace surrounded him and his extraordinarily gentle voice would slur these dictums.

"Sure, sure, we have to fly by the numbers these days. It's the only way to go in an electronic sky because there are only two superb pilots in the world . . . yourself and the last pilot you talked to. But anyone who can swing down from a tree and find their way out of a jungle can be taught to fly by the numbers. You see the product every day and a lot of it should never leave the ground." He grunted as if personally insulted and added, "Such people leave a stain in the sky."

Yet Slonnie was incapable of snobbism, nor did he subscribe to the prevalent airline-pilot hauteur as a standard for his own estimation of other flying men. I have heard him lament, with full complement of rueful gestures, "Hell . . . that man will *never* learn to fly!" Often enough he was referring to a ten- or fifteen-thousand-hour airline veteran, and when I stopped to consider the addressees of Slonnie's invective I could only agree. The typical man he scorned was indeed a pilot, a rough and mechanically thinking driver totally preoccupied with the pay and seniority aspects of his job. Such a man could tell you instantly his bidding prowess according to seniority, the pay details of every run on the system, the exact date of his retirement and the emoluments to be received thereof; but the runways all along his route were dented with his landings and his most elemental aerial maneuvers were erratic and uncertain.

Perhaps you may now entertain certain misgivings about flying people in general, or if you have already captured your share in the world aloft then you may dare examine your status as a pilot or airman. Very well, the gauge is simple enough, and if you will be as ruthless with yourself as you might be with others, the truth is easily read. Does the candidate, first of all, have a passion for flight itself and steadfastly regard it as something more than a means of transportation? Does he lace his devotion with a sense of humility so that he may still and forever learn from his betters and thus not shy from imitating their skills?

Beware of the flying Thespian, he who exhibits upon his person various aeronautical regalia such as big wings, or oversized matching chronometers, or billed caps, which are better suited to protecting the skin of deep-sea tuna men. Alas, these heroes are on stage at many small airports. Turn your eyes away from them unless you have never seen an aircraft wait for the approach of its driver—hanging its head in shame.

If a candidate recognizes the rudder pedals on an aircraft are not installed solely for the purpose of steering it on the earth, then he may be a capable of making perfectly coordinated turns in the air and holding his altitude precisely without reference to any instrument other than God's horizon.

If he really knows his business then he most certainly does not require instruments, horns, bells, and/or whistles to inform him an aircraft is about to stall and he is utterly confident it will never stall unless he deliberately wills it.

Be certain that a true airman has lost his naiveté about our marvelous modern engines and is always aware that they *do* very suddenly disappoint us. He is not dismayed if and when this annoyance occurs.

The true airman despises the long, low, drag approaches which are taught by so many "instructors" and even encouraged by the FAA who secretly know better. "There's less chance of overshooting the runway," they say. Start with a clumsiness and fix with a clumsiness. No true airman would encourage landing with a thump while the aircraft is still in approximate cruising altitude and speed. Even if the candidate is personally incapable of remorse, who gave him the right to embarrass the aircraft?

Honest airmen make their approaches via a glide slope appropriate to the local environment and depend upon final flareout to set their aircraft down with the tender loving care it deserves. Slonnie blesses them from on high.

Slonnie also smiles upon airmen who request a landing "advisory," and holds only contempt for the sort of driver who begs for landing "instructions." The same goes for flying radio "Hams," and so be parsimonious with your monologues and refrain utterly from use of the word "over," or worse, "over and out." Any airman knows such resolute signals are employed only by actors in TV dramas and have been superfluous ever since the introduction of VHF. Air traffic controllers know a true airman when they hear one and particularly when they do *not* hear from him except on request. They do not appreciate a chatty personality who rambles on like a network commentator.

There are certain commandments in the airman's bible, and Slonnie was always a magnificent Moses. He believed you should know how to sideslip, spin at least two turns and recover, do a straight loop without falling out of the top of it, land uphill, downhill, and across hill far from any tower or paved runways, and he pronounced you should know how to work your way through an assembly of cumuli so they will not bite.

He recommended that any aspiring airman fly as often as possible at night, if only because nocturnal good weather can offer some of the most soothing flight sensations available. He specified that any self-reliant airman should be capable of continuing and safely concluding his night flight if suddenly all the lights in the house go out.

These are only a smattering of the many self-inquiries any flying person should try. If your own answers are mostly positive then others should fall into line. You will then be in harmony with the wisdom of Slonnie, and it is likely you are an airman rather than merely a driver.

Once your status is established you may be sure Slonnie would recommend your spending as much time as you can spare at any airport engaging yourself in the business of separating airmen from drivers. Why bother? With his easy smile old number *one* would shake his Indianlike head and slur, "You can always tell when a man has lost his soul to flying. The poor bastard is hopelessly committed to stopping whatever he is doing long enough to look up and make sure the aircraft purring overhead continues on course and does not suddenly fall out of the sky. It is also his bounden duty to watch every aircraft within view take off and land."

In recent times seat-of-the-pants flying has been almost totally rejected and often held in ridicule. I join Slonnie in the belief this is a tragedy because such rejection is based upon the premise that there is greater safety in flying by the numbers, which is in itself a very shady and proofless conception. We need more airmen in the sky and fewer drivers, a goal which can best be achieved by individuals acquiring the skills to employ all the wondrous electronic aids now available to us and basing such abilities on a sound foundation—which should be in the seat of our pants.

An airman nurtured in the best of the old and polished with the numerical disciplines of the new might very well earn a nod of approval from old number *one*.

Chapter Six

The Seventy-Sixth Day

*T*hose who participate in aerial searches nearly always find more than they are looking for. This may be because aircraft have so often gone missing in wilderness areas, and the vast terrain below, unmarked by man, has a way of creating a sense of utter hopelessness in the searchers. The total effect is not immediate, rather it is sneaky like the invisible growth of an interior affliction, and if the search continues for long those involved discover a humility they may never have known before. "How little thou art, how puny you be . . . where in this wilderness aloft and alow is reassurance that I am as important and as capable as I have previously regarded myself?"

Perhaps this yearning for justification is one of the reasons so many aerial searches develop trouble for the searchers. Someone in the crowd, bound for heroism, presses the weather just a few hundred feet beyond his ability to cope or stretches the range of his aircraft into the twilight of its cruising range—and alas, do thereby complicate matters by becoming an object of search. It is an aerial truism beyond dispute that nothing can endanger or frustrate the success of a search quite like a squadron of well-meaning low-time pilots roaring off into the yonder with eyes narrowed for first sight of a wounded aircraft. They simply cannot conceive how very small a clue they are looking for, if only because they have never seen it squashed like a tiny insect against the planet earth. Worse, the mind stoutly rejects the diminutive size of fellow human beings if they have been so foolish as to stray from the wreckage of their plane. Once beheld, the reaction is at first triumphant, a sensation which is almost instantly replaced by awe and abject humility.

Joe Crosson of the newly formed Alaskan Airways, though a true veteran of the north country, knows yet another ordeal during which *his* status as a human being is brought into proper perspective. For this is the *seventy-sixth* day since his partner Carl Eielson and mechanic Earl Borland vanished in the desolation below.

The year 1930 is still very new and the Territory of Alaska still relatively untouched by "Black Tuesday" of October 1929, which in faraway Wall Street triggered a multitude of dire events including the eventual unemployment of more than 16 million Americans. Joe Crosson, whose reputation as a bush pilot is already high, is only mildly interested in an outside world where Thomas Mann has won the Nobel Prize for literature and where mechanics at Valley Stream, Long Island's Curtiss-Wright airport, charge two dollars per hour. Nor does he care a damn about a new name being bandied about a land far to the west of his snow-crusted horizon. After all Mao Tse-Tung is not a man who knows the very special techniques required in flying off skis and floats as well as wheels in a country that sissies say God forgot.

Crosson does not care what people from the outside say about his Alaska. He believes in its majesty and future, even though he is well aware of its terrible cruelties. He is one of a very special breed of men, a tough, elite, extraordinarily resourceful band of flying brothers whose "bust 'em up, fix 'em up, fly 'em out" philosophy has been repeatedly confirmed by deeds. This is the country where the "flight with the serum" stories originated. Crosson himself has delivered relief supplies to Nome during ice blockage of the waterways, and he has flown antitoxin serum to remote villages stricken with typhoid. Like his comrades, Crosson's summer flight uniform consists of a pair of breeches, high laced boots, a sweater, and a rather greasy cap. In winter his haberdasher is an Eskimo woman who knows how to fit skins and keep a man from freezing in sixty-below-zero weather or worse.

Crosson and his tribe do not trouble themselves with formal flight plans but are extremely careful to keep one in their head, and they are masters of flying's greatest maneuver, the 180-degree turn. They leave word at their base when they take off so that when they fly to such regions as the Crazy Mountains, Peril Strait, Cape Suckling, Rainy Pass, or Hog River someone will start things moving if they fail to return as advertised. Yet alarm is not immediate. No one is going to become overly concerned until a flight is at least three days overdue. If something has gone wrong in the bush the chances are the pilot himself can fix it, or if he also carries a flight mechanic then the certainty is even greater. Ingenuity developed of necessity inspires these men to substitute and fabricate, scheme and improvise repairs that would challenge the factory in which their aircraft or engines were made. Paradoxically, many of them are better fixers than flyers. Only a very few have any knowledge whatever of instrument flying, and their comprehension of navigation, aerodynamics, meteorology, density altitude, and weight and balance remain beautifully basic. "If it won't get off the water it's overloaded."

"Turn where the river turns, but when you pass the third lake on your right, turn hard left three minutes later. Then you won't run into the goddamn mountain."

"Compass course? Drift angle? Ever try to fly by a magnetic compass at latitude seventy?"

"Meteorology is the art of anticipation and remembering that in Alaska the weather is consistent Jekyll and Hyde. In two hours a fine soft day can produce a howling blizzard."

Instrument flight is being caught in a "white-out" and knowing just how and where to drop that red burlap bag you've been carrying so you will have something in the dead white pall to give you depth perception when you descend for a landing.

Aerodynamics is appreciation of Alaskan tides, which are the second highest and fastest in the world. If you have floats on your plane and are working the fish canneries do not tarry over-long with an outbound tide.

You may find the ocean half a mile away and your aircraft rendered aerodynamically impotent.

And then there are the gale force winds which trouble everyone except the mosquitoes. You must be hardy enough to snowshoe-pack a runway so you can get away from some remote section of the interior where the temperature may be fixed far below the zero mark. Then once away, with skis for landing gear, you must land in the mud back at Anchorage.

It is a great life if you are willing to risk yours for $400 a month and $4 per hour actual flight pay. Sometimes you collect fares from miners who pay off in raw gold, or if your passengers happen to be trappers a bundle of fine skins may handle the accounting. Or you might be flying on a percentage basis which does cause a man to hump for all the trade he can find.

Flight regulations? There is a rumor about some outfit called the Department of Commerce and Licensing, and all that, but they haven't found their way to Alaska yet. Someone said they would be displeased at your carrying extra fuel in five-gallon cans and feeding it direct to the engine via wobble pump. Maybe it is a makeshift arrangement and a little smelly sometimes, but it's better than going down in the bush.

"Look out for your own ass" has been the standard policy since Roy F. Jones, a salesman for Hills Brothers Coffee, became the first commercial pilot in southeast Alaska way back in 1923.

Joe Crosson loves Alaska, but he is really in love with his Fairchild 71, which he regards as a superior weapon in his defense against the innumerable arctic hazards and a reasonable guarantee of his economic survival. Its load-carrying ability is astounding and that's what counts in the bush. Crosson has hauled everything from prostitutes (sometimes listed on the manifest as "slot machines"), in a country where the five-to-one ratio of men to women makes them valuable freight, to woefully sincere missionaries who are trying to convince Eskimos that a properly welcomed guest is *not* entitled to every hospitality including the host's wife.

Sherman Fairchild originally designed the 71 to carry his camera and mapping gear, and while he was about it created a wonderfully stable airplane with a fuselage of welded steel tubing and a sturdy high-lift wing of box-type spruce spars with wooden ribs. The whole is covered with fabric which is light and easy to repair and the big round Wasp engine has a growl that means business. The Fairchild has a metal prop and even brakes for those times when it is equipped with wheels. The engine has an inertia starter with a booster magneto which can produce its own thrills if you crank it with wet hands.

A very special feature is the Fairchild's wings which fold back after a heavy retaining bolt is removed. In Alaska this ability becomes particularly valuable in riding out gale winds on the ground, and never mind if the operation gives passengers the willi-waws.

Only the speed of the Fairchild has created a modicum of ridicule. With floats it may cruise at 100 miles per hour, but to achieve that speed the load had better be light, which it rarely is. And some say Pacific International Airways, one of Crosson's competitors, had those long red arrows painted along the sides of the fuselage so people would know which way their Fairchilds are flying.

Competition between the various outfits which have come to Alaska is cutthroat, but when a plane is down or lost a truce is declared and everyone in the flying business joins the search. On that terrible day in early November when Eielson and Borland climbed into their Hamilton "metalplane" the winds were gusting to fifty miles an hour and by noon the temperature had managed to warm up to only fifty below zero. They took off because they had a fat contract, the kind that can make the long winter a little easier to bear.

A schooner named the *Nanuk* is locked in the ice near a place called Chigitum in Siberia. She is loaded with furs which must get to market. Crosson and Eielson's Alaskan Airways has the contract to complete the job a sailing vessel is now unable to do.

It was soon known that Eielson and Borland did make it to the schooner, took on a load, and began their long flight across the arctic wilderness for Teller in Alaska. They never arrived. Now, over two months later, Crosson refuses to give up. Somewhere in the vastness below his friends might still be alive and waiting.

Crosson is a determined man and long established as one of the breed. He has shared the Alaskan skies with such bush-flying royalty as Monsen,

CROSSON'S QUEST
Fairchild 71 over the Alaskan wilderness

Mills, Clayton Scott, Noel Wien and his three brothers, Percy Hubbard, Bob Reeve, Frank Dorbandt, Harold Gillam, and Frank Barr.

Among other achievements Crosson made the first commercial flight between Fairbanks and Point Barrow in the very far north and was the first over Mount McKinley at 20,320 feet, which is very high for these times. He is now chief pilot for Alaskan Airways, and as if to confirm the Crosson family's place in northern flying, his sister Marvel has become the first woman pilot in Alaska.

Yet Crosson's impressive record, his hard-won skills, and even his fierce resolve have thus far produced nothing but growing despair. There has been no sign of any human disturbance to the frozen wilderness below.

Crosson does not depend on his own weary eyes to find a clue, for there are many others searching different areas. Among them is Gillam who was also engaged in bringing the furs from the *Nanuk*. There is "Pat" Reid of Canada, also flying a Fairchild, and even a Russian, Commander Slipnev, flying a single-engined Junkers on skis. They have flown to the limits of their fuel endurance, through Christmas 1929 and then New Year's, which normally they would have celebrated with a liberal application of whiskey and maybe a song.

But you do not sing when one of your kind is down.

Here is yet one more reminder that Alaska and the arctic are far from being conquered. Flying these skies in winter is partly a business of convincing yourself that sweet fortune is your most faithful friend.

At long last, when even Crosson has nearly lost his resolve, when the combination of bitter cold and persistent flying have brought him near exhaustion, something flashes in the everlasting bleakness below. Or is it, Crosson wonders as has every other aerial searcher, just something his imagination sees?

He turns, seeking a repetition, perhaps even a denial of his rising hope. And sure enough, there it is again, not a flash, but at least a glint of sunlight on what appears to be the tip of a metal wing projecting from the deep snow.

Crosson rocks his wings to signal Gillam who is nearby in another Fairchild. They make one pass to judge the terrain and confirm the fact that they have actually been lucky enough to stumble upon a pinpoint in infinity. It is the seventy-sixth day of their searching, and now there is no question that their primary task is done.

They land, leave their engines ticking over, and flounder through the heavy snow toward the wreckage. Panting from their exertions they dig down and clear one of the cockpit windows. Inside there is no sign of Eielson or Borland.

The more they study the interior of the Hamilton and more convinced they become that the one-time occupants are long gone. Yet in this frozen wilderness they could not have gone far even in the best of weather.

Why did they go down? Where are they?

Crosson remains with the wreckage while Gillam flies back to Teller, Alaska, for help in the final stages of the search. The crash site is in Siberia, but international frontiers sometimes have a way of becoming conveniently invisible to airmen of every nation. Soon there are men enough on the site, including the amiable Commander Slipnev and his Junkers, its right wing buttressed Eskimo-style with ice blocks to serve as a tolerable shelter for the digging crews.

Finally, some twenty-one days after Crosson first spied the wing tip in the sun, the diggers come upon Borland's body. Five days later their fanlike series of trenches, which extend outward from the wreckage, reveal the hard frozen body of Carl Eielson.

One hundred and two days from the first report that they were missing, the bodies are flown to Teller, one in Crosson's Fairchild Yukon and the other in Commander Slipnev's Junkers. The American flags which draped the bodies of two rugged individuals who so obviously believed their government was their servant rather than the other way around were made by the Russians.

There really isn't much to flying the Alaskan bush. All you need is a good pocketknife with many blades (can opener, screwdriver, punch), some emergency rations, and a notebook to remember all those things people in the boondocks ask you to bring them "next trip." And a typical working day might include flying 300 miles with the mail, two round trips to a gold mine in the interior, stopping by another mine to pick up an injured man, fish-trap patrol for a few hours, then off to the western islands to drop off a cannery official. Somewhere en route you should have time to munch on a chocolate bar. In Alaska you make money flying while there is light enough to see, and in the summer that is as much as twenty-four hours a day. And if you are lucky, you will be flying a Fairchild 71.

The important part, winter or summer, is the spirit within yourself which *must* match the work at hand. Communications are rudimentary, but Bob Ellis, a pilot for Alaska Southern, expressed that spirit best when he sent a message from Ketchikan. He had experienced considerable aggravation with his engine which had been rude enough to quit on takeoff. After a bush-style fix to the engine, bad weather delayed a second departure. Fearing his headquarters in Juneau might start a search for him, Ellis radioed his situation while being careful to stay within the ten-word limit.

"Waves high. Should engine die, goodbye. I are ready. Ellis."

Chapter Seven

The Newsboy

When the stuff of hangar flying becomes thick enough to muddy all belief then the time has come in western coastal lands to tell of the Columbia River Gorge and why it is so very much deeper than it used to be. Then you will learn it was not alone the mighty river coursing its way down to the sea, but hard-pressed airmail pilots trying to sneak through the mountains from west to east who enlarged and deepened the cut. The frequently inclement weather obliged them to fly so low the blast from their prop wash eventually eroded the sand and stone.

This touching legend of getting the mail through may hopefully be repeated to Gorge visitors in the years after 2000. The antiseptic citizens of tomorrow, who probably will never know the special stink of hazardous work, may be hungry for tales of derring-do.

In 1932 United Air Lines pilot Joe Smith has reached the venerable age of twenty-three.* The great Tex Rankin taught him to fly, and he has been at it four years, mostly over a route which (always excepting Alaska), has the worst flying weather in the United States. Because of the high terrain it may thus be ranked along with the worst in the world, not except-

---

* Captain Joe Smith retired with 31,000 hours logged flight time.

ing northern Europe where at least the majority of clouds are not stuffed with granite. The route between Oakland, California, and Seattle, officially designated Airmail Route 8, customarily produces every known meteorological phenomena. There is fog to match any along the East Coast, winds exceeded only by the Alaskan breed, rain as a way of life, and ice to match the cruel brand found along the Mason-Dixon line. If anything is lacking to challenge a man then the United pilots who fly the run that was once Pacific Air Transport's and Boeing's are grateful for the relative paucity of thunderstorms.

Among Joe Smith's comrades on the route are Virden, Miller, Tyler, Laughlin, and Cunningham, names that will eventually identify senior royalty in the world's largest airline. None of them ever thought they would be called "Captain." Dammit, a pilot was a *pilot* and the stars on his sleeve, which designated a thousand hours for each, were enough frill.

Pilots of the western skies came to United via one of its predecessors and they all remember when a certain Verne Gorst formed Pacific Air Transport and paid them partly in cash, which was in extremely short supply, and partly in P.A.T. stock. They held the stock only worthy of papering the little crew house which was hard by Medford airport. There, sometimes torn between frustration and disappointment in their reward they shot at such fancy wallpaper with their airmail 45s.*

Yet even after the merger with United the mail pilots of AM-8 share other mundane problems. Any clerk aged twenty-five can buy $10,000 worth of life insurance for $151 per year. A mail pilot of identical age is obliged to pay $401 *if* he can find an insurer. Home, Metropolitan, and Mutual, among most of the major concerns, flatly refuse to insure any man who flies "regularly."

Otherwise things could be worse. Although the rest of the nation is aching through the curse of the Great Depression, United is somehow meeting its payroll and even expanding. The Boeing 40-B-4 inherited from Boeing Air Transport is a fine flying machine, a man's airplane, aging now, but still nearly without fault. It is ideally suited for the route, and as presently equipped with the new Boeing radiotelephone system, letdowns

---

* When P.A.T. was absorbed by United the stock was honored, bullet holes and all.

through the overcast create a much less severe pucker factor. An artful dispatcher standing on the hangar roof at either Oakland, Medford, or Seattle can judge your whereabouts from the purr of your Wasp engine and direct you accordingly.

"Turn due east and fly one minute, Joe. . . . Okay, back to the east one minute. . . . got you. Turn to three hundred degrees now, Joe, and start your descent. Okay? Level off at one thousand. . . . You're coming fine, Joe. Now steady on two hundred eighty degrees and descend. . . . You should break out about five hundred feet, Joe, with a good two miles visibility underneath. Got ya! Welcome."

But flying needle, ball, and airspeed is not common in the 40-B-4s or anything else. Once in a while a man is caught in solid cloud, and then he sweats it out so literally he doesn't need any additional heat in the cockpit. Or he fails to find a break, in which sad event large red cards are dropped along the route by searching comrades.

IMPORTANT

An airmail plane has been lost in this vicinity. Anyone seeing or hearing a plane in this district during the hours of 8 A.M. to 12 noon Thursday morning, January 22nd, please telephone _____.

It's a long route, and shrewd pilots avoid bucking through on instruments. The best technique is to get right down on the ground if necessary—and stay there. One of the Boeing 40-B-4's few faults is its relative blindness while taxiing, but in the air, visibility from the cockpit is satisfactory providing you know what you are looking for. The fundamental truth to safe "contact" flying in bad weather is the ability to identify every stream, patch of forest, ranch, and railway junction—*instantly*. Just hoping that hamlet which is momentarily visible straight down might be "White Pigeon" or "Black Angus" is asking for trouble. Convincing yourself it is such a place simply because it is due to appear means the beginning of your end if it happens to be some other hamlet. Later, while you are milling around entirely surrounded by bewilderment you would do well to review your sins because they are soon to be reckoned. The cockpit of a mail plane is no place for a man who would lie to himself.

Joe Smith and his comrades are acutely aware of the international human proneness toward wishful thinking. Like another United pilot named Jeppesen they guard against uncertainty by keeping a special little book in the knee pocket of their flying suits. Each pilot is his own author, and he records certain facts about the route in his own way.

AIRWAY BEACONS: ELEVATION
36b Marigold . . . . .2500′
37a Stagecoach . . . . .2500′
37b Cowcreek . . . . .1490′
38 Canyon Mountain . . . . .3000′

The book also contains personally drafted little diagrams of the new radio-range facilities which have just been installed at a few of the major way points. The "legs" are drawn, the *As* and *Ns* marked, and the best courses to intersect a leg are indicated. Another page lists the average flight times between checkpoints, and another lists emergency strips with their lengths and elevations. All of these notations have been meticulously compiled by the pilots themselves. Who else would do it?

Yet by far the most important aids to avigation are stored in the memories of the pilots. In the same fashion river pilots once knew every turn and eccentricity of a river, they know old Macdonald's farm from Hansen's spread. Although they may catch only a fragment of Mosby Creek they will not confuse it with an unnamed creek to the northeast if only because there was no smell of sawmill smoke, and they immediately recognize Lake Fornicate because they christened it so. If they spy the loom of Mt. Adams there and Huckleberry Mountain puncturing the overcast off there, then the course *must* be "that way."

They are friends with the land beneath their wings.

Later in the flight, down California way, even the calendar is verified, for the dependable Mrs. Post who lives exactly eight miles north of Red Bluff Airport invariably hangs out her laundry on Saturdays. There are many other friends below, usually quite unaware of their usefulness, but friends all the same. And when something happens to change the scene, a death, a sale, a remodeling, then it is soon known to the pilots of AM-8.

All such air-ground friendships are not one way, even as the northwest weather is not perpetually rotten.

The summer sun is still high at five in the afternoon, when pilot Joe Smith bound from Portland to Medford, Oregon, usually passes the fire look-out station on Quartz Mountain. During a previous trip Smith had noticed something flashing from the same station and he thought, "That man is lonely." So he left his regular course and buzzed the station. The ranger's enthusiastic waving was enough to convince him all was well. Yet the thought of his isolation stayed with Smith, and the next trip he resolved to alleviate it.

By now a custom has been established and a new friendship formed. Joe Smith buys a newspaper in Portland and soon afterward dives his 40-B-4 at the little clearing on Quartz Mountain. Tossing the paper this afternoon he has risen higher than usual in the cockpit. And one of the buckles on his helmet is loose. As the paper makes a descending arc for the target his goggles are torn away by the slipstream. A futile reach and Smith knows they are gone forever. It is going to be a long flight to Medford without their protection.

Ten days later Smith's mail includes a mysterious package. It proves to be his lost goggles, which are not even scratched.

"I found them hanging in the branch of a tree—"

The Boeing 40-B-4 was not only an efficient and air-kindly craft; it arrived on the aviation scene at just the right time less a year. The original design known as the Model 40A was inspired in 1925 by the post office department, who needed a replacement for the venerable Liberty-powered DH-4s. The post office bought one, and the design was shelved.

Approximately one year later, with the marvelous caprice of so many government agencies, the post office suddenly advertised the transcontinental airmail routes for private bid. Boeing Air Transport, heedless of dire financial warnings, won AM-8 and revived their own 40A design to do the job. To meet the operational deadline, twenty-four aircraft were built in six months. That the work was well done happened to be a very good thing for Boeing because the advent of the type also marked a monumental change in the U.S. aircraft industry. Hitherto there had been no govern-

THE NEWSBOY
Boeing 40-B-4 over fire lookout tower in Oregon

ment control over the construction of aircraft. As of January 1, 1927, the Department of Commerce took over strict regulation of airworthiness and registration. The Boeing 40A model was furnished with Approved Type Certificate (ATC) Number 2.

While they were at it the Boeing engineers decided to hedge their gamble and make room for two paying passengers in a small cabin forward of the cockpit. New type designation—the Boeing 40-B-2. Later the cabin was enlarged to accommodate four people who were obliged to be cozy whether they were inclined or not. Thus the famous 40-B-4.

The basic Model 40 design seemed blessed from the very start. The type went directly into service without the usual lengthy ceremonies involved in testing a prototype. The fine Pratt and Whitney "Wasp" engine had just become available for the earlier models and soon afterward the beautiful "Hornet" engine, which in itself contributed mightily to the progress of American commercial aviation.

Overall construction was welded steel tubing, dural, and fabric covering. A total of thirty-nine were built, the last in 1932. It had such attributes as a tailwheel, three-bladed propeller, hand-operated wheel brakes, two-position rudder bar, and an airspeed and altitude indicator for the interest of the passengers.

The quality of construction may be solemnly weighed against a certain shoddiness to be found in some latter-day aircraft. *Two* Canadian-built 40-B-4s survived to fly in World War II! Pressed into military service by the New Zealand government, both were lost in action during the New Guinea campaign. Two others still exist, one in the Ford Museum, Dearborn, Michigan, and the other at Rosenwald Museum, Chicago.

Chapter Eight

Metal, Gas, and Goldbeaters' Skin

*T*wenty-second-century historians may well be puzzled if in studying microfilms which have survived the vicissitudes of time they discover a strange gap in the progress of transport through the lower levels of the earth's envelope. They will find among other bewilderments that an organization known as DELAG (an understandable acronym for the sneezable "Deutsche Luftschiffahrts Aktien-Gesselschaft") was carrying paying passengers as early as 1913 on more or less regular schedules. By the outbreak of World War I this enterprise had lofted more than ten thousand satisfied customers over Germany without so much as inflicting a bruise.

DELAG knew some horrifying moments during its first efforts, but its nearly incredible luck remained so constant that the German military who had originally taken a sour view of lighter-than-air ships were persuaded against their lingering doubts to build a great many. They were the nucleus for the 161 rigid aircraft ultimately constructed, of which 119 were built by the Zeppelin Company.

Today, in all the world, only a planned community of upper-middle-class stucco houses known as "Zeppelinheim" survives the era. Situated hard by Frankfurt airport, the community is populated mostly by Lufthansa employees and their families. All are aware of the community's heritage, and many know how a certain German count, Ferdinand von Zeppelin, a rather unorthodox cavalry officer, was finally invited to depart the military "for the good of the service." The year was 1890, and like all military establishments since the first rock was thrown, the count's superiors did not applaud soldiers who marched to their own drummer. And like so many other independent souls bound to make either history or the brig the count refused to become invisible. He soon got himself into further mischief by inventing a type of lighter-than-air vehicle which would be identified by his name forever.

Count von Zeppelin was as much an energetic promoter as he was a dreamer, and his first airships revealed something of both characteristics. It was not until the count discovered a misanthrope, an engineer named Dürr, that the many technical problems involved in constructing and flying such enormous gasbags approached practical solution. Quite as providentially, there arrived on the scene a certain Dr. Hugo Eckener, then employed as a reporter for the prestigious *Frankfurter Zeitung*. He knew nothing whatsoever about lighter-than-air and his original opinions on the efficiency and future of Count Zeppelin's earliest designs were anything but enthusiastic. Yet somehow, perhaps because Eckener was also a fierce individualist and capable of giant-sized dreams, the two men developed an affinity for each other. Before Eckener realized what was happening to him he had engaged his literary talents on behalf of Count von Zeppelin, his marvelous device, and his ambitions for German aerial supremacy. The combination of the two men was bombastic and unrelenting, and in their zeal a new and unique cause was born.

By continuous and brilliant persuasion in government and private circles plus a side ante from the Kaiser, they managed to fund the construction of LZ-2 and LZ-3. The LZ-3 made its maiden flight in October of 1906 and simultaneously lifted itself from the suspect ranks of grandiose folly to a tangible Teutonic symbol. (The upturned faces of innumerable proud Germans found it much easier to sing *Deutschland Über Alles* when aloft they had such a magnificent sausage to look at.)

Yet another and still enduring force was set in motion by the early Zeppelin flights. Converts to the theory of lighter-than-air flight came easily, if only because the only competition in mankind's long nourished dream of flight were such entrepreneurs as the Wright Brothers, Bleriot, Santos Dumont, and their far less impressive machines. A further, nearly unexplainable development, an emotional sometimes wildly irrational passion which has survived even into the jet age suddenly took charge of people involved in lighter-than-air. Perhaps the very difficulties of lighter-than-air flight are responsible for the unswerving devotion of airshipmen, or it could be that the aerial behemoths were so visually impressive they became living beings and that any man who had much to do with their operation became secretly awed at his own participation. Whatever the Freudian complex, airshipmen have long been fanatically religious in their aerial belief and have repeatedly demonstrated unswerving faith in spite of disasters. From pioneers Zeppelin and Eckener through Strasser, Ernst Lehmann, von Schiller, Hans Flemming, and the American naval officers Rosendahl, Moffet, and Lansdowne, the zealots have never officially surrendered, and it is inconceivable that those very few in the world who are still engaged in lighter-than-air operations* will ever break the trust.

The First World War tore the *Gemütlichkeit* from the Zeppelins and assigned them to missions which Count von Zeppelin had long used as fund raisers from the military, yet which made him extremely unhappy when his cards of doom were finally called. More dreamer than soldier, he was distressed at the historic bombings of London and consoled himself with the thought that relatively little damage resulted. Raids were also conducted over various targets along the eastern front including the Ploesti oil fields, where during World War II the USAF would experience its own tragic "charge of the Light Brigade." The loss of men during the whole Zeppelin effort was out of all proportion to the military value, but as happens in all wars, some residue of benefit sifted through the carnage. When peace was declared the Germans had over a thousand men trained in lighter-than-air operations, and they knew better than any other people how to build the only vehicle capable of intercontinental transoceanic flight.

In July 1928 Countess von Brandenstein-Zeppelin, only daughter of

---

* The Goodyear Blimp crews.

the great man himself, christened LZ-127 *Graf Zeppelin*. The gods were smiling on all lighter-than-air champions that day, for eventually the *Graf* set a series of records never to be matched by other airships and did more to justify her existence than any of her kind.

1935. The Zeppelin service to South America is firmly established with the *Graf* making her fifteenth trip of the year. Once again *Gemütlichkeit* has become the tone aboard airships and with one curious exception the passenger accommodations aboard the *Graf* are designed to enhance that atmosphere. There is a lounge in the gondola which serves as dining room and social center. The tasteful curtains and carpets of burgundy red compliment the carefully selected wine list. The food is excellent German fare albeit on the heavy side as befits ladies and gentlemen who can afford Zeppelin tickets. Aft there are sleeping cabins for twenty passengers, four to a cabin, but there even such relatively simple luxuries end. Washrooms and toilets are provided but actual bathing is out of the question. Water weighs too much for any quantity to be carried, and even the waste water is kept aboard to avoid a compensating valving of hydrogen to cover its loss. And it may be some measurement of total preoccupation with the intricacies of airship control and flight that no provision whatever has been made for heat in the gondola.

It is a time of international financial stress triggered by the collapse of the American stock market, and a time of dictators in too many countries. While arch-villain Adolph Schickelgruber is waiting in the wings, Mussolini's voice echoes from the Italian boot to the Alps, Ataturk struts in Turkey, Stalin is letting Russian blood, Trujillo has just taken the Dominican Republic to his cast-iron heart, and Antonio de Oliveira Salazar is admonishing the Portuguese to cease their rude habit of expectorating wherever and whenever they please. Many conservative Americans complain that they are ruled by the most vicious and conspiratorial dictator of all—Franklin D. Roosevelt. Yet in the midst of such discouraging images Dr. Hugo Eckener and his *Graf* have managed to establish a proud record.

Soon after her first 1928 testing flights, the *Graf* rose into the milky sky over Friedrichshafen and pointed her great snout westward—destination Lakehurst, New Jersey, U.S.A. Eckener, who was certainly one of the very first airmen to understand and utilize the "pressure" system of oceanic flying, had little stomach for a head-on encounter with the North Atlantic during October. He knew the seasonal westerlies could very well net him a groundspeed near zero if not an actual minus. So he headed south for Madeira and the Azores and, except for some hairy hours punching through the line squalls which so often hunker along the Azorean horizon, Eckener finally managed a triumphant arrival. The occasion was considerably aggrandized by the "emergency" which Eckener had declared en route and which found its due place in American headlines. Overconfidence in his mighty ship had influenced Eckener to carry on at full speed through a particularly choleric squall with consequent damage to the *Graf*. Fabric was torn from the horizontal stabilizer and threatened unpredictable and extremely dangerous aerodynamic behavior of the ship unless immediate repairs were made. Eckener brought the *Graf* to a stop while volunteers, including his son, climbed out on the fin and fixed her wounds. Finally, the SOS to the U.S. Navy was canceled, and the airship proceeded on her way having now captured the complete attention of the American public— which was exactly what Eckener sought in the first place. No man to be hit on the head twice with Thor's hammer, Eckener rang down a cautious "slow ahead" on his engine telegraphs when they entered a second series of squalls off Bermuda, and no damage resulted.

Despite the success of the early flights which included a magnificently executed voyage around the world in 1929, the *Graf* was never quite at home over the North Atlantic. She did find the perfect environment over the South Atlantic and by 1935 had made fifty flights to Recife in Brazil, many of which continued on to Rio de Janeiro.

Now with the South Atlantic service so successfully regular as to approach the humdrum, the operational limitations of ordinary aircraft combined with the highly specialized business of lighter-than-air flight have placed the Germans eras ahead of all long-range competitions. England, the United States, and France, all having bloodied their own LTA record, try to swallow their collective doubts as the *Graf* continues her remarkable career.

Yet what should be a total vindication of Zeppelin travel has been unexpectedly soured by German home politics, and the outspoken Eckener himself, who has been openly critical of his Führer, would have been elimi-

THE MIGHTY GRAF
Dirigible Z-127, Graf Zeppelin, cruising over Rio de Janiero harbor

nated if he were not such an international hero. The Zeppelin show, however, is no longer his personal affair. Air Minister Goering, in unusual and bizarre cooperation with Propaganda Minister Goebbels, has taken over via the *Deutsche Zeppelin Ruderei*, a government-controlled company. To Eckener's dismay and chagrin they have caused a huge Nazi swastika to be painted on the *Graf's* port vertical tail surface while the German tricolor is allowed to remain on the starboard. Among other antagonisms generated by such gross display is the flat refusal of helium gas for German Zeppelins by the world's only supplier — the United States government. Perpetually harassed, Eckener can only console himself with the continued performance of the *Graf* and his preoccupation with his ideal airship to be christened *Hindenburg*.

The now venerable *Graf* is powered by five Maybach engines, which burn "Blau" gas, a fuel similar to propane, and is hoisted aloft by hydrogen which is hardly noted for stability. Eckener and his disciples have been flying one of the most potentially explosive assemblies ever united. And yet nothing serious has happened. The good and lucky *Graf* has only been indirectly involved in earthly events such as the outbreak of a revolution in Pernambuco, which the *Graf's* Captain Lehmann handled nicely as far as his vessel was concerned. He simply drifted off the South American coast for three days until advised by wireless that hostilities had ceased and the government troops were once again in charge at the aerodrome. Then he rang down "full ahead" on all five engines and made a normal landing at Recife.

Merely holding altitude and position for such length of time is clear demonstration of highly polished lighter-than-air technique. For flying a dirigible or any other sort of gasbag is an endeavor only distantly related to winged flight. Even the principles are but second cousins. In the *Graf*, as in all dirigibles, there is an "elevator man" who stands facing thwartships. By activating a large wheel he changes the pitch of the ship in much the same way an airplane is controlled with a stick. There the similarity ends since the kinetic energies involved bear no comparison. A "phugoid" effect is generated as a result of variations in the longitudinal movement of the airship's center of mass in flight. While an airplane responds immediately to any adjustment of its control surfaces, dirigibles, as all lighter-than-air ships, take their own sweet time and the nose of the *Graf* may be actually rising while the elevator man has spun his wheel to full-descent position.

Likewise the "rudder man" who stands at a second wheel in the bow of the control car must master the art of anticipated delay. He may be given the order for a 20-degree turn to the right and spin his wheel in his response. The *Graf* is 787 feet long with a buxom 138 feet of beam. Such mass does not exactly pirouette under the hand of any mere man, and when the nose does start to swing around the horizon it is not inclined to stop even near a pre-selected target unless the rudder man has developed a very alert sensitivity to his charge's habits and is simultaneously blessed with smooth air and a sympathetic commander. Flying any lighter-than-air ship is akin to riding a whale over which the passenger has relatively little control. Landing is so complex and subject to uncertainties that *Graf* commanders like to have a ground assist from at least three hundred men.

There are further grace notes and vagaries to the demanding orchestration of Zeppelin flight, and even a considerable herd of cows contributes to the ultimate performance. The nearly 4 million cubic feet of hydrogen gas which gives the *Graf* her lift is contained in 17 gigantic cotton bags lined with goldbeaters' skin—the filmy outer membrane of a cow's cecum. It is gas-tight stuff and light of weight, but the financial penalty is stunning. The *Graf* requires 850,000 skins to keep the hydrogen confined where it is needed, which even by Texas standards is a lot of cattle off the range.

Controlling the location of the *Graf*'s hydrogen is only part of the problem. As night follows day just as surely will the *Graf*'s lift be influenced by the difference between the outside temperature and that of the gas. Known as "superheating" and "supercooling" according to whether the gas expands or contracts in response to temperature, either condition requires careful attention by a Zeppelin crew. If expansion of the hydrogen carries the ship above a desired altitude, gas must be valved, which is like drawing money from the bank. If the gas cools, with a consequent loss of lift, ballast must be dropped, which is like throwing money out the window.

In Los Angeles, where a temperature inversion is often present, the *Graf* arrived over Mines Field at dawn after crossing the Pacific. Descent for landing became an outsized dilemma. At 1,600 feet the temperature was

77 degrees, below it was 66 degrees, a cool bath in which the *Graf* refused to submerge until vast amounts of hydrogen had been valved. On departure the following night the problem reversed itself and the *Graf* refused to ascend until ballast plus six crew members were off-loaded. Only then did Eckener dare ring down "full speed ahead" and employ dynamic lift to become airborne. During the ensuing struggle for altitude Eckener maintained his dignified calm with difficulty including some very anxious moments when the *Graf* cleared the high-tension line which then as now so often seems to be standard airport fencing equipment.

Graf crews normally listed 45 to 50 men. Her useful lift was 66,000 pounds and her cruising speed 73 miles per hour. While it is inconceivable that she ever operated at an honest profit, her consistency paid off in reputation for her builders and operators. The *Graf* had flown around the world and was regularly flying the oceans before Americans could fly across their own country in even the slowest of airplanes. While the rest of the world had no commercial air transport at all or at best flew quite primitive aircraft over very short distances the *Graf* celebrated 144 ocean crossings and the safe transport of over 13,000 passengers.

In her almost 600 flights the *Graf* filled a now easily forgotten aeronautical gap between the world wars, and even her retirement after nine years of service seemed insultingly inappropriate to her superb record. After the *Hindenburg* disaster her thirteen flights planned for that year were canceled. She made a final and faultless ascent and flight between Friedrichshafen and Rhein-Main Airport at Frankfurt, where she was put on display to the public. Her engagement as a museum was short-lived. In the spring of 1940 she was dismantled, and various of her pieces were allocated to the German war machine.

The skies will never see her like again.

Chapter Nine

The Flying Brooklyn Bridges

American soldiers returning from Siberian service after World War I sang, "The wind she choke in Vladivostok," and in Buffalo, New York, winter 1936, they will find themselves quite at home. Or any winter. For here, the mighty winds of middle America sweep eastward across the expanse of frozen Lake Erie and howl across Buffalo airport making the working day of American Airways mechanics one long, benumbed ordeal. But times are tough and the men are tough.

American business has only recently emerged from the storm cellar and the Dow-Jones has climbed to 160. Congress has passed an act called Social Security, but there are still 9 million unemployed in the land. And in spite of the lingering disillusionments of the Depression most Americans still believe in honor, work, country. So to hell with the chill factor.

For the airlines in the United States times are almost as grim as the winter gales, and there are still echoes of collusion in the awarding of air-mail contracts during the present administration. As a result most airlines will soon or have already changed their corporate names—slightly. The New Deal has not been their dearest friend.

Passengers are just beginning to discover a few extra dollars in their purse and are filling about half the available seats. On good days.

AN AMERICAN SIBERIA
Curtiss-Wright T-32 Condors at Buffalo, New York, airport

Still a passenger favorite is the remarkable Curtiss Condor, which now, after only a few years of operation, is fighting for its existence against more than winter tempests. Here is an aircraft which broke tradition and arrived on the scene at exactly the right time. And yet promises only a short domestic career. The stalwart trimotored Fords, some still in service with most of the major airlines, are too slow, uncomfortable, and carry only twelve passengers. After a brief relation which turned out to be little more than a flirtation, three- and four-engined Fokker transports have been found too inefficient and expensive to carry passengers from Glendale, California, to eastern cities where the overwhelming majority of Americans live, especially those likely to have money enough for air fare.

Precisely when most needed this civilian version of a military bomber became available to the airlines. Eastern Airlines and Transcontinental Air Transport powered their earlier versions with 12-cylinder Curtiss Conqueror water cooled engines. American Airways has chosen the Wright cyclones and are glad of it, particularly in winter weather. The reduction of engines from three or four to merely a pair has cut operation costs until the Condor has the lowest per-seat mile in the business. Instead of fretting about the supposed safety margin to be enjoyed with a third engine, passengers are delighted to be rid of its heat, smell, and noise. The decibel count in a Condor is 75, about the same as a Pullman car. A trimotored Ford clangs along at 105 decibels, a number which almost exactly matches its cruising speed.

In this winter of '36 you can still fly in a Condor Newark-Chicago in 5½ hours, or Newark-Washington in 1½ hours.

Pilots have developed an almost maudlin affection for Condors. Cockpit visibility is superb, acres of sky better than the view from a Ford. Condors land like a baby carriage and are stable enough to keep the captain, who wears two gold stripes on his sleeve, in a mellow mood. In decent weather such is his content that he may even give his copilot, who wears one and a half stripes, the next landing—*providing* he goes back and helps the stewardess serve the box lunches. The first box, in obedience to a process of natural osmosis, will somehow find its way to the captain's lap.

There is one great curse upon the Condor which not even the master of all pilots can remove. It has a nickname, the "flying Brooklyn Bridge," a snide appraisal grumbled most emphatically during winter. This ridicule is inspired by the considerable wing area and the more than few struts and flying wires necessary to keep the wing panels of any large biplane flying in close formation.

The sound of a Condor on final glide to earth may be an inspiring symphony to air buffs and romantics, but the pandemonium attendant to a Condor caught with a load of ice is enough to give the boldest pilot serious bowel trouble.

All the great men who fly the Condors during these days are graduates of the open-cockpit–helmet-and-goggles school. Many famous names are among them—Dick Merrill, Tom Boyd, Ernie Dryer, Benny Howard— none are men of fragile nerves. Yet recently the pucker power of two American Airways stalwarts has been severely tested after their Condor acquired such an elaborate decoration of ice that an immediate landing became imperative. They chose an open field and thanks to the Condor's benign landing character plus skillful piloting, no one was injured and there was no damage to the aircraft.

While passengers and crew were still congratulating themselves, a choleric policeman appeared and demanded explanation of their unauthorized trespass. Unfortunately, the answer included dialogue spiced with air-foil deterioration and lift coefficient which sounded like a frivolous reply to the constable. And so with ominous tone and a snarl about "wise guys" he took out his summons book and inquired the names of the crew. The captain offered his name as Daniel Boone, which the officer accepted with growing suspicion and further comment on smart mouths. When the copilot gave his name as Kit Carson only the intervening pleas of the passengers kept him from hauling both pilots off to gaol.*

Respected she may be, but everyone in aviation knows that the Condor and all her buxomy kind are doomed. For two years, American Airways, Colonial, Eastern, and TWA have been flying a sleek Douglas called the DC-2. Though it carries only fourteen passengers in comparison with the Condor's eighteen it does so faster and in better style. And it can be bought sparkling new for a reasonable $65,000.

---

* They were giving their true names.

Even more serious competition now threatens the Condors. In the flying business success has always been spelled BIGGER, and so the DC-2 is already being altered to become something called a DC-3. And the Eskimo telegraph transmitting right off the frozen ramp at Buffalo airport has it that American Airways will fly its first sleeper version to the West Coast in the fall of this year.

Chapter Ten

Category 1930s

*D*uring these sweet times of Omni navigation, DME, radar, and Category II coupled approaches, there is rarely occasion for long-weathered pelicans to employ certain survival tricks learned as fledglings. This is good, and I am very grateful that aviators no longer need to be part gambler plus two parts adventurer and employ the cunning of a Machiavelli mixed with the determination of a commando to complete an instrument flight in one piece. A man or a Ms. can now embark on an instrument flight, and while the yawning quota may become minimal the scalp remains pleasantly loose from takeoff to final. The sense of knowing precisely what is transpiring *all* the time allows you to fly in an aura of total well-being unless the weather is extraordinarily nasty.

It was not always thus. There was a time when computers were something you put in your shirt pocket, flight plans were handwritten even on the airlines, and winds were quoted in miles per hour. Knots were for sailors.

Nearly all instrument flying in the United States was accomplished by the *domestic* airlines in aircraft that accommodated as many as fourteen passengers plus one stewardess to ease their ill-concealed fears with chewing gum and a smile.

During these same times I knew a young lady who dwelled in a house some miles east of Cleveland Airport. Just before hailing the tower and warning of our DC-2's imminent approach I would make a steep descent thereby advising the young lady that our arrival matched the printed schedule which she had obligingly memorized. Regardless of visibility her house was always easy to find. It was smack on the easterly leg of the Cleveland beam, and she had caused a large white cross to be painted on her roof. Carried away with enthusiasm for our appearance she would sometimes wave a sheet from an upper window. During the winter when our arrival was after dark she would bundle up and stand in the street waving a powerful gasoline lantern.

I cannot remember how many of these not-so-gentle rendezvous occurred, but I am certain that because of the rather steep banks involved the passengers could see the girl in the street as well as I could. Yet they never complained. I could only conclude they must have thought it all part of the Cleveland instrument-approach system.

Actually the semaphoric lass *was* an aid to navigation. Having spied her I knew exactly how many miles we were from Cleveland Airport. She was early model DME.

The 1930s are already old before I have at last achieved enough hours to win a right seat with a major airline. Like the majority of applicants, I have arrived without an instrument rating simply because few nonmilitary pilots can afford the very expensive training. Several of us are enrolled in the airline school to learn the art, and our instructors regard us with misgivings. We are at the bottom of the airline pecking order and possibly untouchables. If we turn out badly we may even break their rice bowls for there has been considerable complaint from the regular line captains about the folly of wasting money on a *school*. "The place to learn instrument flight is in the cockpit goddammit!"

Since veteran captains often have the ear of officials, the instructors are making very sure they do not turn out an inferior product. The lash is frequent and heavy, our eyes smart with tears of perspiration as we labor endlessly in a device known as a Link trainer and alternately one of the several Wright-powered Stinsons the airline employs for pilot route checking of emergency landing fields.

These are the days of sweeping transition in the flying business, although we are all much too preoccupied with the relentless severity of our instructors to appreciate anything beyond the accusing instrument assembly of needle, ball, and airspeed, plus the familiar altimeter, rate of climb, and two wonderful luxuries few of us could ever afford, an artificial horizon and directional gyro. Like my student comrades I am often confused and worried about squeaking through the course without disaster, a promise we find in every instrument flight, simulated or real.

Primarily we are in terror of becoming lost, a situation which most of us have previously experienced many times, but never with some critic sourly witnessing our performance and certainly not when visibility stops at the engine cowling. Maintaining our bearings and sensory wits is not made any easier by the instructors' cruel notion that we should be able to fly instruments with one or several of them inoperative. They carry little black patches which nicely fit over each instrument. These are applied one at a time until we are left with only a magnetic compass and altimeter—the latter only if we have, after the initial shock of losing the airspeed instrument, switched to "alternate source."

"Now," the instructor announces with the compassion of a doctor advising you that you have a hideous disease, "you have picked up a dandy load of ice. Make the following turns and no skidding into them."

The man is a sweetheart. For now you notice that he has at least left the ball of the bank-and-turn instrument barely visible below the lower edge of its cover. His reasoning is that it would not be affected by ice.

Your opinion of his gentility soon curdles for as he calls off the various 90- and 180-degree turns to the left and right and then various courses to fly, including several which involve northerly turning error on the magnetic compass, you gradually become aware of an ominous silence in your headphones. Where is the sound of bacon frying which is so much a part of radios in summertime, and more importantly, where is the monotonous cooing of the Chicago radio range?

"And now," declares your dear instructor, "you have been flying various courses for ten minutes during which time the weather has made beam reception impossible. But you are a lucky fellow and you are out of the bad stuff. So here is your radio coming back again. Find yourself. You have five minutes."

You try to peek out from the side of the hood but it is just not possible. You receive an irritating drop of sweat in your eye for your trouble and wonder if you can somehow sneak the hood up on your forehead one-eighth of an inch. In seconds it strikes you that the instructor would express himself most pithily if he caught you. Seconds later you realize that perhaps someday, in that glorious millennium when you command an airliner, there will not be a hood to push up when your position is unknown or any other salvation except your own skill. So you initiate one of the several ways you have been studying to relieve the most sickening of sensations, that which is a feverish nausea known only to a lost pilot under time pressure. Now, depending on where you *think* you might be you can resort to the "true fade" system or the "90-degree orientation" to find your way out of limbo and into the smiles of your instructor.

It was in this fashion that many young men, who had rather wistfully left helmet and goggles to mildew in a closet, did eventually learn what the big boys were doing. True instrument flight for general aviation was simply not done, if only because the four-legged radio range could be tricky and capricious and very often in bad weather, when you needed it most, it became unreadable. Yet in instrument weather, radio ranges and very occasional fan markers were the *only* navigational aids provided for normal operations. Somewhere in our manuals there was a loose statement that in the event of dire emergency and the failing of every other effort we might try calling for DF bearings to be taken upon us, which would provide a reliable steer. The manual neglected to state where in the continental land mass beneath us properly equipped DF stations might be situated, even if their staff might be versed in taking useful bearings. None of us had ever heard of an actual direct assist from the ground.

As a consequence those who repeatedly flew instruments soon developed great resourcefulness and self-reliance. And if you knew the tricks and gave it total concentration when the signals were difficult to read, the four-legged range became a very loved device. In time we learned from each other, the chieftains passing on lore to the younger men at their side.

I have flown with men who could hear many useful signals when I could hear nothing but crescendos and crashes of static. They were like a skillful conductor catching a sour note in the brass or, in the midst of a soaring symphony, lifting an eyebrow at a millisecond's lateness in the timpani section of his orchestra. Reading a radio range in bad weather was more an art than science.

Our own maestros were of various temperaments. Some liked their beam turned up to high volume, which had its faults, while others kept it tuned down past the point of apparent audibility—which also had its evils unless you were that particular maestro.

These were times of independence when some men lowered their seats all the way down and turned the cockpit light to full intensity in a thunderstorm and others just sat there and spit right back at the lightning. All of them knew the wintertime brought climb, cruise, and descent without seeing the sun for weeks at a time and that often the ice which accumulated soon after takeoff was still there when they landed.

Now, all radio ranges in the United States have been silenced forever with their duties taken over by the visual and ever so much more dependable Omnis. Loss of the ranges is certainly not mourned by those who flew them, but their tricks and individual eccentricities are still remembered by many pilots who have long since accepted their retirement watch.

In spite of government efforts to make it behave, the Salt Lake Range was notorious for its "multiples" and certainly contributed its share to a long series of tragic crashes. You might be sitting at 12,000 feet, fat and happy with the mellow note of the Salt Lake Range telling you how very fine you flew. The two identifying signals would be of equal intensity, and as if that were not enough, the proper A and N signals would be mixed to the exact formula you preferred, which would be a strong "on the leg" with a very slight "feathering" of the right side of the leg signal. (Only clumsy pilots held to the center of the leg during initial approach.)

You are soothed by this dangerous lullaby and totally untroubled in your world aloft, ". . . listen . . . am I not a master of this business? Hear how I remain so firmly just in the twilight zone, inbound to the cone of silence. I have, with consummate skill, so nicely bracketed the very edge of the leg it has been unnecessary to change course more than a few degrees in spite of the wind."

Few pilots who regularly flew the Salt Lake Range allowed themselves such secret and premature congratulations. For they knew that on occasion it could purr as faithfully as any other in the land and then inexplicably turn to claw at them. The least consequence was worry and fright. The worst lured them into the surrounding mountains. The wary knew that one minute on the Salt Lake Range your headphones would be singing a smooth "on course" and the next minute a bold and confusing *A* or *N* would suggest you were flying in a wide open quadrant.

"Where the hell did the leg go? I had it pinned down."

Anger, fear, or just dismay depended on the reported ceiling and visibility at Salt Lake Airport. United and Western pilots, cursed with two of the worst ranges in the nation, Salt Lake and Burbank, were regarded with special respect by those of us who flew better behaved stations in the Eastern flatlands.

Yet nearly all ranges had some idiosyncrasies. One might have a bend in its northwest leg which could cause you concern about the wind if you were unaware of it. Another, during certain conditions of the weather, offered you a number of "on course" signals, each near enough to the true leg to confuse and sometimes confound the listener. Some ranges would fade as you approached or build in volume so suddenly you became convinced that your carefully worked out ETA over the cone was several minutes late.

All route intersections and many holding patterns were based on the crossing of two legs from two different ranges. Since one leg or even both legs might be subject to "swinging" through an arc of several degrees, the points of intersections were far from precise. In thunderstorm weather, when you yearned for clear signals to help you twist, connive, and slip successfully between the towering cumulus, trying to gather a few range signals could become sheer physical torture. Many aviation doctors are convinced the marked deterioration of hearing acuity in older pilots is not so much due to age as to the punishment their ears took from combined engine, slipstream, and especially radio range noise.

Although thunderstorm conditions created havoc with all low-frequency radio systems including communications, the most frustrating and sometimes worrisome situations were brought about by snow static. It was quite unpredictable.

Beyond the cockpit windows, a few inches beyond your own nose and that of your DC-2's, lies the night. Range signals are crisp, the air smooth enough to drink the stewardess's lukewarm coffee without fear of spilling it. The company communications frequency is also clear. All is very well in your cramped little world aloft, and to heighten the cheer you have turned all cockpit lights to full intensity. Matters are so nicely in hand you might even flip through a magazine while the copilot improves his instrument proficiency. (Nice fellow. He has the wisdom to realize you are a virtuoso.)

To better concentrate upon the contents of the latest *Saturday Evening Post* you have pushed your headphones away from your ears, but not so far that you might miss a call on company frequency.

Suddenly you are aware the copilot is shifting unhappily in his seat. His eyes are questioning. "I've lost the range. Nothing."

You deposit the *Saturday Evening Post* in the aluminum bin which already holds the metal logbook and skid your headphones back in place. Nothing? Listen my child and you shall hear—?

You must know what you are listening for. Alas, the child with only one and a half stripes on his sleeve has spoken the truth. There are no signals of any kind or the rasp of distant voices from anywhere in the night below. There is only a gentle hissing in your headphones as if some wag were playing a recording of ocean waves singing on a beach.

You reach for a switch above your head and flip on the landing lights. Suspicion confirmed. Out of the night trillions of white lines are lancing toward your eyes. Snow. Apparently the finer the flakes the more effective. It has isolated you and all aboard from the nether world. The total effect suggests you might have become a passenger in Captain Nemo's fancy submarine.

A call to the company brings no response. Suspicion reconfirmed. You will now proceed via a combination of dead reckoning and flight plan until at least some fragment of a signal manages to sneak through the snow static. Nothing can be done about it, and the situation may endure for five more minutes or an hour. The temptation is to reach once more for the magazine—nonchalantly. Have you not been similarly tried many times before, and has it not always been that eventually *something* will be heard?

You are grateful even if all is not as serene on this flight as you had hoped. You have plunged into the snowstorm while in level cruise with still a few hundred miles to go before descent. Your flight plan is on record with ATC, and they will be moving the small block of wood which represents your aircraft in accordance with the airway and your corrected groundspeed. They would like a few intersection reports which are obtained secondhand via your company frequency, but not having received any word of your progress they *suppose* you are just about where you are *supposed* to be.

Without flipping too far back in your logbook you can revive certain over-long periods when snow static or the thunderstorm variety created a much more nerve-wracking situation.

There would be that time holding over Newark, which was still serving as the airport for New York City. Aircraft are stacked up to ten thousand, all milling around, and waiting for descent. There is no common frequency. Each aircraft works ATC through its own company frequency; thus if you are flying for American you can only guess the whereabouts of the United flight which took off from Chicago just before you and was also bound for Newark. Who made the best time? Is he behind or ahead of you? Or did he land in Cleveland? And then somewhere up here are two TWA flights and a Colonial Airways arrived from Montreal. That is—maybe, if you are to believe the last communication with your company. And there just *may* be a military or itinerant aircraft in the vicinity which *may* be reporting directly to the range station voice facility—or may not bother. So far there has never been a collision on instruments, but how big is the sky? The speculation lies in fragmentary communications separated by moments of total loss between all concerned.

You gather clues like an apprehensive detective, while making sure you are at least holding your last assigned altitude. In an attempt to roughly verify your position you crank the direction finder to and fro and you record bearings on commercial broadcast stations WEAF and WOR. They do come through with some distorted Tommy Dorsey, and the bearings will at least keep you from straying too far over the Atlantic Ocean while you wait for something to happen.

For the past twenty minutes, while bouncing around at seven thousand feet, you have managed well enough with this sort of information. Your fuel-versus-endurance calculations plus some concern over what may have happened to the weather at your alternate, Baltimore, has been spaced by interminable periods of stony silence from below.

"American flight . . . (garbled)"

Was Newark calling *you* or one of the other two American flights in the area? You wait, which is the smartest thing to do until things clarify. Suddenly they do, however slightly. That call had nothing to do with any of your tribe now holding over Newark. It was Memphis Operations querying one of the new sleepers on its way to Burbank. Now again—

"American flight six clear to . . . (garbled) . . . after passing . . . (garbled) . . . Newark altimeter twenty-nine eighty-nine . . . (garbled)"

Silence. You have the altimeter anyway. You wait for a moment, switch on the left landing light for a look at the snow as if your frown of displeasure would make it go away. Then you call Newark for a repeat. Response is immediate, and this time the puzzle looks better.

. . . (garbled) . . . (garbled) . . . (garbled) . . . you are clear to garble five thou-garble and cross Metuchen inter-garble . . . five garble. After passing garble garble three thousand on garble west leg of Newark range . . . expect final approach garb-lance at eight-garble garble. Traffic is garble at garble-thou. . . . United at four thousand . . . TWA at garble garble garble."

"Roger." Piecing together the important fragments and adding local knowledge to custom translates into the following: "You are clear to descend to five thousand while proceeding to Metuchen intersection." Easy enough if you can hear both the Allentown and Newark range legs which create that invisible joint in the murk known as "Metuchen." You may have to guess a little, but after you're convinced you had passed you can drop down to three thousand on the southwest leg of the Newark range, which is conveniently the approach leg. As for the alleged time of final approach it is relatively unimportant since ATC often misses such estimates by fifteen or more minutes. There is very little you can do about the garbled traffic altitudes even if they had come through quickly. They have already changed. You just hope ATC has all their wooden blocks and altitude slips in the right slots. Considering the ever-increasing amount of traffic entering the New York area, ATC is doing an outstanding job. Only a very few flights are lost in the shuffle, and no one has been hurt.

Some notification for the passengers is now in order. While you make the copilot's night by allowing him to execute this initial letdown and see that he is aimed in the general direction of Metuchen intersection, you take pen in hand. In the kit box at your left side there is a handy form for this sort of thing, and you fill the blanks rapidly. You record the flight number, date, flight from Chicago to Newark. Next to the time you write altitude as *descending* and, with some misgivings, an estimate of the time your wheels will scrape the cinders of Newark Airport. You lie ten miles an hour about your airspeed in case some passenger might compare notes with that of another airline, and under "Remarks" you add graciously, "Nice to have you aboard."

Below this comforting phrase the company has added its printed epilogue. *Please pass this slip to the passenger in back of you. Your stewardess will be glad to provide you with a copy of this report if requested.*

You sign your name and forge the copilot's after the fancy title "First Officer." Then you click a switch at your side three times to call the stewardess and note with mixed emotions that while at this lower altitude the Newark range is becoming ever more readable, the company operations remains incommunicado. Since you are now down to three thousand you would appreciate some further information on the present ceiling and visibility at Newark.

Fortunately, the company frequency at Memphis is still coming in loud and clear, and so you give them a try. Response is immediate, which is the way of things on low frequency.

After a few "standbys," Memphis, which is a thousand miles from your position, relays a final approach clearance. Maybe it would be easier if the dispatcher down below would fold his paper work into a paper airplane and sail it up to you.

You are sliding down the Allentown southeast leg and beginning to pick up bits of the Newark range. You sense the stewardess stumbling around in the dark passage behind you which contains some of the passenger baggage and the steam heater. (The balance of the baggage plus a thousand pounds of sand in bags fills a compartment just forward of the tail. For something called weight and balance they say.)

You have been listening to those occasional notes of the Newark range which manage to penetrate the static wall, and you are content that during the past seven minutes you have been flying a reasonably direct course for Metuchen intersection. After resetting the directional compass too many times you decide that there is really not much of a north wind, but rather that the device itself is precessing too rapidly. You squawk it in the metal logbook. Even while writing you maintain an old and valuable habit based on years of self-reliance. Without seeming to nursemaid your copilot you make damn sure he holds altitude, course, and airspeed, and also that he keeps the carburetor air temperatures in the safe range and avoids cowboy turns. *Fifteen-degree banks, my friend. Company regulations.*

Of course, if *you* are making the turns, 45-degree banks are quite acceptable, particularly when the passengers can't see anything.

Even though your hands are not on the controls you are in effect flying this aircraft in your mind, and therefore it is quite just that you should log the instrument time.

An aroma of perfume mixed with an ever so slight hint of feminine worry informs you that the stewardess is about to step to your side.

You move one headphone a trifle so you can hear her.

"When will we land?"

Ah, little sweet-smelling registered nurse, how innocent your question. When you worked in the hospital did you ask the doctor when a patient could go home? "Soon," he might say, or "Never." The holding times for us are becoming ever longer, and we may wonder where all the passengers of the future are going to be found. Did you know, girl, that they are tearing up a place called North Beach Airport, which is hard by Flushing, and Mayor LaGuardia declares it will be replaced by a monster aerodrome of the future?

"How many passengers are back there?"

"Nine."

"Have them make a landing pool. Anytime within the next forty minutes might win."

You hand her the passenger information slip and ask if she would like to stop by Harry's for a drink on the way home tonight.

"Thank you. My boy friend is picking me up." A giggle.

No matter how mighty you may be up here there are inevitably certain things that will trim a man back to size.

Now at Metuchen it is back and forth and around and around with the engines throttled back to conserve fuel, but not too far back to render carburetor heat ineffective. You pass the time by removing the holster from your side which contains the .38 caliber automatic which post office regulations specify shall be worn by any person who handles the mail aloft or alow. The copilot is similarly armed, but no one in all of the airlines has ever had cause to employ the weapons even as a scare device. Not too long ago a drunken baseball player went berserk in the cabin of a DC-3, but the crew subdued him with a vigorous application of a fire extinguisher to his iron head. The guns are a nuisance, and when you must walk all the way back through the cabin to relieve yourself people ask why you wear them.

"Afraid of bandits?"

"Well . . . (a smile) you never know."

Then when you are finished explaining the pistol you are in for a barrage of other questions. And there is always the veteran of World War I who insists on telling you how he became an ace.

You place the gun in your flight briefcase, which is just behind your seat. It is pregnant with operations manuals, avigation charts for the whole radius of your possible cruising area, the Civil Air Regulations, a manual on the DC-2, a book of open railroad tickets in case you are grounded somewhere and have to "train" your passengers, two flashlights, and several personal items.

You pull out the instrument-approach plate for Newark, which is printed on typewriter-size paper. It shows the four legs of the Newark range and their magnetic courses. Little airplanes are portrayed on each leg with the minimum instrument height for that leg inscribed across the top of their wings. The actual approach is detailed graphically by more little airplanes flying the descent courses after passing the cone of silence on initial approach or direct approach from an outer fan marker which is rarely authorized.

Obviously, from your present position you will proceed directly in on the southwest leg, pass over the cone of silence at three thousand, make a 180-degree turn, descend outbound again on the southwest leg, make a procedure turn when you feel you have everything pinned down, then return to the cone of silence still descending. One minute after passing through the cone, if the ground is not in sight and you have reached minimums, pull up and either ask for another try or get the hell along to your alternate.

The minimums are specified as 300 feet and ¾-mile visibility, a fact with which you are as familiar as you are with all the other details on the approach plate. And who is going to monitor your personal minimums at Newark or any other place along your route? No matter what the operations office specifies, the "peek-a-boo" minimums still apply and each man has his own unless a chief pilot happens to be riding the jump seat. For example, on a night like this, sneaking into Newark should be easier than on a hazy summer night with twice the ceiling if only because some aeronautically naive groundlings decided to cover both Newark, and Washington, *and* Boston airports with black cinders. Naive, or did the politicians have relatives in the cinder business? There are no runways, which is helpful in crosswinds, but the first view of black nothing after breaking out of a low overcast at night does keep a man intensely interested. At Boston it is often impossible to isolate the black airport from the black water surrounding it, and at Newark the marshes to the south and north also melt into one nothing with the airport. By the time you pick up and identify the perimeter lights it can be too late. Given a night like this with snow covering the cinders there will be something for your landing lights to reflect upon. The final flare-out should be easy.

While you are debating whether or not to make the supreme gesture by giving the approach to your copilot, Newark Operations suddenly comes through with the latest weather.

"... garble garble ... three garble ... vizgarble one garble ... light blowing ... garble ..."

To no one's surprise this translates as ceiling three hundred (you know the numeral three does not mean three thousand) and visibility one mile in light blowing snow; which just happen to be the official minimums and therefore permit you to descend for a peek. It all depends on who makes the observation. One man's mile is not necessarily another man's mile.

If after descent you dislike the look of things you are free to depart, but then what if another flight from your own company makes it? Or worse, supposing you decide you can't hack it and United or TWA does? And if there is an Eastern flight anywhere in-bound they will certainly slip in. They are particularly competitive at Washington, which is also black cinders—if that miserable ex-sandbank beside the Potomac can really be considered an airport.

You place the approach plate back in your bag because you have long ago committed it to memory. Reception is improving. The intermittent tone of the Newark range is rising to a high whine and is beginning to dominate the bacon frying in your headphones. It becomes a reasonable certainty the cone itself must be only a few miles ahead.

You are in a magnanimous mood in spite of the stewardess's rejection of your after-duty charm. Alas, it had been your scheme to ease *her* into the same generous spirit by telling her of nights now long gone when you flew for an audience of stars and navigated by the winking lights of airway beacons stretching fifty miles ahead.

So you allow the copilot this initial approach and watch his tightly pressed lips while he blows it. Nearing the cone his anxieties overwhelm him. He over-corrects and there is never any true silence. The built-in signal strength occurs on schedule, but instead of the significant silence followed by a sharp renewal of signal strength, there is a mixture of signals, A's and N's and the identifier N-K all mixed and rapidly replacing each other. But at least he has come close enough to the cone to catch a marker light on the instrument panel. The signals are fading, and so his miss was not too disgraceful.

Yet for his delinquency he must suffer. You place your hands on the controls. "Not too bad. But you can do with a bit more seasoning."

Just to reassure your own secret self you take up a course corresponding to the southwest leg and turn the radio volume down until the signals are nearly inaudible. They fade very rapidly. The station is certainly behind and you have only to find the southwest leg which is buried somewhere in the frying pan. Close-in reception is not at all bad, and so after two minutes you have the on-course nailed down and continue the descent. You proceed outbound while easing over to the east side of the leg, and know you are there when you start picking up a faint N.

Now then, three minutes outbound according to the book, then a procedure turn to the east. Thanks, you'll take four minutes and buy an extra minute to bracket the leg inbound.

At two thousand feet, with the procedure turn completed, you pull the window open at your side and remove the gum which has served as a pacifier for the past hour. You hold the gum delicately between thumb and forefinger, then position it just inside the window edge. You open your fingers and release the gum. It vanishes instantly, sucked into the roaring slipstream as if yanked away by an invisible string. It is a petty amusement signifying nothing except that you now intend to settle down to serious work, but at least the gesture is more civilized than the not uncommon practice of disposing paper napkins, Kleenex, banana peels, and even whole lunch boxes via the same convenient route. Apparently no one who lives in the depths below has ever complained of the litter which must decorate the earth below all air routes. Or is there just a great deal of uninhabited America? Perhaps those below, who only a few years ago endured bombardment via the open-hole toilets in tri-motored Fords, consider anything an improvement.

Spearmint gone, you slam the window, and the hollow roaring subsides. On the last part of the procedure turn you call for the gear down and slide into the final approach leg. You continue descending at 400 feet per minute while eyeing the sweep second hand of the panel clock.

You ease into the center of the leg and have no trouble hanging on to it although it becomes increasingly narrow. As you feel your way along the acoustical tightrope you find the wind is westerly instead of northerly as forecast. What else is new?

Your private guess is one minute and thirty seconds to the cone of silence—make it one and twenty-five seconds because of the wind.

At the one-minute mark and now down to 1,500 feet, you call for 10 degrees of flaps. The copilot pumps mightily on a long lever at his side and for his exertions watches a small white metal arrow move along a track beneath the instrument panel. When it is approximately in position he covers his flashlight with his fingers, then parts them just enough to allow a small beam of light to seek out the arrow. Good lad. He has been carefully taught. Many captains become short tempered when flashlight beams are waved about indiscriminately during the final approach. They claim it spoils their concentration.

The important music has now begun, and the noises which so often distort the dulcet tones of a reliable range are muted. You listen most carefully for a hint of an *A* or *N* which would warn of your straying from the center of the leg. Popular magazines refer to this as "riding the beam" although you have never heard any pilot describe the mixture of signals which creates this steady tone as anything but a leg.

The copilot quickly releases his seat belt and rises. With less than a minute before final approach his action is displeasing.

"Where are you going?"

"I forgot to turn off the heater."

"Leave it be."

The copilot resumes his seat. Mixed with the sound of the range you hear the clinking of his seat belt buckle. The heater is a steam contraption circa late Robert Fulton, and no great harm will be done if it is left on until you touch down. You do not confess that you also forgot about the heater. Times are changing so fast. Could it be there is too much going on during final for a man to remember? Or two men? The operations manual includes a check list which details what should be done before landing, but who is going to haul that heavy book from his kit, riffle through a hundred odd pages and read off what you already know? Three green lights and suffi-cient hydraulic pressure are all a man needs to tell him the gear is down and the flaps are ready when needed. There is talk of handy check lists you are supposed to read at a glance, but so far it is all talk. Thankfully. Fighting your way out of a paper blizzard is bound to be harder than this approach.

You have cut off the company frequency so you can concentrate on the range signals. Just as the cone of silence begins to build, when you least need distraction, your headphones rattle with your final clearance from ATC. The copilot repeats it.

"The company says we are cleared to land. And Eastern just missed their approach."

Hmm-mm . . . Eastern Airlines' pilots have a long-established reputation for pushing weather. They are very good and very tough to follow, and one who suddenly decides he has had enough is cause for worry. Perhaps the Eastern pilot is new to the area, or is long on fuel and temporarily short of adrenalin. Or just feeling liverish. No matter. You are now committed for at least a peek.

The on-course signal is building very sharply. You turn the volume down to save your ears and estimate fifteen seconds to the cone.

"Give me another bite of flaps."

The copilot pumps dutifully, and you shove the two position props to low pitch. Speed is 120.

Now a subtle game begins, for it is your business to display an air of nonchalance in spite of any subsurface emotions that may be gnawing at your own adrenalin pipes. Copilots move from one captain to another and what they have seen in you to admire goes with them. And sometimes what they do not admire. Wise copilots are not gossips, but capturing and holding the respect of another man is a fundamental masculine need. In the tight little world of professional flying there are in the whole nation less than a thousand pilots who fly instruments regularly, and fewer still who shoot 300-foot approaches in blowing snow at night. Therefore, when the peanut light glows to indicate the cone and the signal crescendos and almost instantly falls away to complete silence, it behooves you to fake a yawn.

"Give me full flaps."

As the copilot pumps, you pull back on the throttles and shove the nose down. The range signals are now reversed with the *N*s and *A*s changed sides. As you leave the station the signal volume diminishes as rapidly as it mounted. The marker light goes out. You have allowed your speed to drop off to 105.

Because the distance from the range station to Newark Airport is so short you are only half-listening to the signals of the outgoing leg. Your attention is divided between four demands. First the altimeter which is unwinding rapidly. Without altitude you are permanently out of business. The rule has always been, "Careful preservation of your own ass will likewise insure that of each passenger."

The next preoccupation is time, and so you watch the sweep second hand with great interest. One minute from cone to airport under no-wind conditions. All right, tonight allow six seconds extra for a possible north wind. Wise men know that if you forget the exact time you passed through the cone or fail to watch the clock there is a potential for trouble. You may reach minimum-instrument altitude too soon and start looking outside when your attention should be strictly inside. Or you may let down too late and miss the field entirely, or worse, catch a glimpse of the ground and be tempted to circle by eyeball; which every wise man knows is a very bad way to stay in business.

Holding an absolutely straight course is another secret linked to a good approach from the cone. There is always a compulsive yearning to chase signals, particularly if there is a bend in the leg or multiple courses. The mandate is keep that turn needle straight up and down and be damned to outside influences. You want the airport in the front windows not out on either side.

Oddly enough the most aerodynamically important factor is now worthy of only an occasional glance. Airspeed has become merely a complement to time. In training and checks you have made many approaches with the airspeed instrument covered. You can feel this airplane with the seat of your pants. If it begins to feel sick you'll know it without the help of an instrument, and when you need it the most, as in ice, the chances are you won't have it.

Finally there is the artificial horizon, which is a nice aid for holding attitude. It is like a doting aunt—nice to have around but not essential. That nice, old married couple, needle and ball, still maintain their dignity.

Twenty seconds to go and 105 miles per hour, which is about as slow as you care to proceed until you break out underneath—if you do. Altitude 400 feet and still descending.

"Sing out when you see the ground."

The copilot nods and loosens his tie. No fear about his failing to look out. He has his nose so close to the front window his breathing makes a small opaque circle on the cold glass. You are keeping your own eyes in the cockpit where they belong.

Ten seconds and 300 feet.

Without taking your eyes from the course, instruments, and clock, you feel above your head for the right-landing light switch.

"See anything?"

"Not yet." Strange. This copilot's voice is cracking. Are they hiring boy scouts these days?

You hold 300 feet for another five seconds just to keep things honest. And now comes the human factor which mixed with local conditions ultimately decides the practical minimum descent altitude for any airport.

With snow on the ground at Newark you are willing to slip down to 250, maybe, if there are a few breaks, down to 200. But not for long because there is a formidable psychological blockade to the north of the airport. It is a very tall chimney possessed of magical powers which allow it to move closer to the airport in direct proportion to the tightness of ceiling and visibility. Because all the terrain about Newark is flat and the immediate vicinity of the airport free of hazards (unlike Chicago which has various high tanks, powerhouse stacks, and tall buildings on every one of its four range legs), you might hold two hundred feet for ten seconds, but no more.

Once having broken out, scud or dubious visibility may demand that you keep your eyes outside. With clock watching no longer possible you must sometimes resort to aviation's oldest method of keeping time in seconds—raising your tongue to the roof of your mouth, which during such moments is always quite dry, bringing it back into place again and counting. Each cycle is one second. It is more accurate than the equally venerable one-thousand-and-one, one-thousand-and-two method.

You halt the altimeter at 250 feet . . . and start counting.

"I have the perimeter lights."

You glance out ahead and immediately flip on the landing lights. Two bright lozenges appear in the snow. They seem to be racing each other. It is all over and you have only cheated by fifty feet.

The perimeter lights slip past as you flare very slightly to make a wheel landing which is most pleasant and satisfying upon a cushion of snow. "Are we down?"

Later, taxiing into the American terminal, a small, one-story, shedlike building, you ask the copilot if he happened to notice the exact altitude you broke out of the overcast.

"Three hundred feet."

Good boy. He knows his official minimums, and his discreetness will carry him far. Now if he will just remember to open his window and put up the flag without being reminded—

In 1930 the aeronautics branch of the Department of Commerce installed an experimental instrument-landing system at College Park, Maryland. It consisted of three elements, a runway localizer, outer and inner marker beacons, and a landing beam or glide path. The Army Air Corps installed a similar system at its own Mitchell Field, Long Island. Both systems worked, yet by the end of the 1930s and even into the 1940s the radio range system remained the only method of instrument navigation available to U.S. pilots.

Under the sluggish hands of any state, fulfilling the needs of man invariably takes a very long time.

Chapter Eleven

Saint-Exupéry Country

*D*uring the year 1926 an American named Robert Goddard launched the first successful liquid-fuel rocket thereby commencing a long series of experiments which would eventually send men into space. Of far more importance to Frenchmen was the same year's production of some excellent burgundies although the clarets could only be denounced as almost devoid of nose and therefore were regrettable. For Frenchmen it was an imperfect year in other ways. In May a national hero was believed lost, none other than Jean Mermoz, aviator par excellence, who had been forced down in the Mauretanian desert and captured by Moorish tribesmen. Having distinguished himself in combat along the western front in World War I, Mermoz, like so many demobilized pilots of every other country, resolved to continue his career in the air. He managed this determination by becoming one of the very first pilots of the Latécoère Company, which in itself was destined to become not only the most glamorous airline in the world but also a man killer.

After a failure of Mermoz's engine brought him like a wounded falcon into the arms of the waiting Moors they debated killing him for the sheer pleasure of demonstrating that a white-skinned European had no business flying over their desert domain, but the more practical leaders decided he would be worth more alive than dead and named an outrageous ransom, which for the continued glory of French aviation was duly paid. Soon after his release, Mermoz was back flying the French and European mails over the same territory, his élan typical of the men who flew for the Latécoère Company.

Today in that Parisian oasis, the Bois de Boulogne, situated over a fine restaurant, there is an exclusive club known as the Vielles Tiges. Its most revered members are those Frenchmen who flew when Nieuports and Spads were their mounts, and there are still a handful who were flying prior to the First World War. They are old now and crotchety as only elderly Frenchmen can be, and unless you speak their language flawlessly, foreign visitors are initially welcomed with total indifference or greeted with the enthusiasm of disturbed porcupines. But in time they relent, their latent charm and camaraderie asserts itself. If you are of the flying breed and will listen they will allow their eyes to wander fondly over the priceless collection of photos and trophies which adorn their lair, and they will expound most generously upon those exciting times when Frenchmen pioneered the air routes over the Sahara and down the west coast of Africa, thence across the Atlantic to Argentina and beyond. If you are very lucky even the great Jean Dabry might be present. Still looking no more than sixty, vigorous, bright of eye, and possessed of a mischievous smile, he was Mermoz's navigator on Aeropostale's first flight across the South Atlantic which in a sense represented the culmination of all previous efforts to establish a regular airmail service to South America.

Habitués of the Club Vielles Tiges will discuss almost anything as long as it is aeronautically oriented, and the names they drop so casually are the stuff of French air legends and the principal well from which a certain poet-pilot known as Saint-Exupéry drew forth his masterful tales. They will speak of places, with an aura of adventure in the very enunciation—Cape Juby, Villa Cisneros, Cape Bojador, Agadir, Port Etienne, and simultaneously they will identify desert tribes good and bad, the Isarguin, the R'Guibat, the Ait Gout, and the fierce Ait Oussa. Men they will name, now legendary characters who flew the route from Toulouse to Senegal and eventually all the way to Patagonia and Chile.

The roll call is likewise suggestive of the full life, a dangerous, proud, and hectic existence as led by such stalwarts as Paul Vachet, Beppo de Massimi, Reine, the doughty Guillaumet, the seemingly indomitable Mermoz, and Saint-Exupéry, whose fame would long outlast his comrades.

The magic of Saint-Exupéry's pen carried another French early bird to immortality, one who might otherwise have become entombed in the clammy archives of French bureaucracy or, at best, vaguely remembered as having had something to do with the final debacle of Aéropostale and the attendant scandals. He was Didier Daurat, the original Operations Manager of the Latécoère Company, which in time became Aéropostale. He was the hard, unforgiving yet sensitive "Patron" of the line, and there is very little question that Saint-Exupéry used him as a model for the lonely, disconsolate Riviere in his poetic novel of the Aéropostale's Argentinian endeavors, *Night Flight*.

Daurat was the "essence" of Aéropostale, its spirit and its inspiration. He realized from the beginning that the prospect of flying the mail from France to South America was a fearful undertaking, especially since available men were not trained for the job and the machines they must fly were ill matched to the severe environment.

Daurat was a realist: "A pilot who marries loses three-quarters of his value here." Yet even Daurat could not foresee that in ten years' time "the line" would claim the lives of 121 Frenchmen.

"If a pilot is wrong then . . . just below the sea of clouds begins eternity."

The French aircraft manufacturer Pierre Latécoère, who had made his fortune building Salmson biplanes during the war, had the nerve to propose flying mail to North Africa and perhaps South America even before the guns of 1918 had ceased their murderous thundering. When the armistice was declared he hastened to establish the Latécoère Airline Company and appointed Daurat, a twenty-nine-year-old veteran *poilu* who had managed to shake the mud of the trenches and finish the war in the air, as his operations chief.

Daurat began his task with characteristic thoroughness. The "line" became his life, as it had to be since the odds against success were obviously overwhelming.

Latécoère said, "The experts claim what we intend cannot be done. *Voilà!* We must do it."

Because of their limited range the Latécoère airplanes must cross the Spanish peninsula if they were to achieve the North African coast. The German-oriented Spanish authorities proved so uncooperative that for the first years Daurat's communication with Spanish bases was via carrier pigeon. Downed pilots regardless of cause were arrested by the *Guardia Civil* and the mail confiscated. It was only a foretaste of an attitude which would even further harass Daurat and the southbound operations of "the line" when they were obliged to fly over the Spanish Sahara. And although the Germans had been defeated along the western front the German colonies in South America remained extremely vigorous and powerful and would eventually give Latécoère, and later Aéropostale, even more trouble.

Fortunately, two quite exceptional men engaged themselves in a tricky campaign to smooth the way for "the line" and at least allow it a fighting chance. The urbane Neopolitan, Beppo de Massimi, eventually persuaded the Spaniards to display some semblance of neutrality; and in South America the fabulous Vicente Almendoz de Almonacid (Croix de Guerre with palms, Legion of Honor, etc.) insisted his fellow Argentinians should give the French a chance. He went so far as to challenge one of his fellow countrymen to a duel when he refused to cooperate.

These diplomatic skirmishes were only a part of Daurat's continuing battle. Piloting aircraft was not yet a profession, and those who knew anything at all about flying were combat veterans who were not overly amenable to discipline and were less than enthusiastic about flying in the sedate fashion Daurat wanted them to fly. They thought it absurdly bourgeois to fly straight and level, and to forsake a few acrobatics on arrival, wherever it might be, was to emasculate a flying man. If the weather was bad then the way to improve it was through generous applications of cognac to the pilot, who during the recent hostilities had become quite accustomed to facing death on every flight.

His rambunctious brood nearly drove Daurat out of his wits until he resigned himself to loneliness and began running things with an iron fist. Pilots who broke his rules were promptly fired regardless of promises to reform. New pilot hirelings, among them the great Mermoz, were first put to work with the mechanics.

"It may humble them," thought Daurat, "and perhaps they will learn something."

The "line" was fairly well established before Daurat, with considerable misgivings, hired a tall, young man who said his name was Antoine de Saint-Exupéry. His manner was diffident, his physical coordination seemed awkward, and he was low on flying time as well, but when in spite of his aristocratic background he cheerfully went to work with the mechanics and stayed with them uncomplainingly, Daurat became encouraged. Eventually he worked him into the regular flying schedule. Thus, in time, some very fine literature was born.

Other unforeseen dividends resulted from Daurat's democratic program. Pilots and mechanics came to understand and sympathize with each other, and soon a magnificent *esprit* was established throughout the organization. Without it Latécoère, and later Aéropostale, might not have survived the grim years of struggle against the hostile routes and the international turmoil generated by their very existence.

During the earliest days of the Latécoère Company the Breguet 14 was used all along the route as far as Dakar in Senegal. It was a war surplus biplane of mixed construction—duralumin and steel frame covered with fabric. The first aircraft put into service wore their wartime mottled camouflage and military numbers until they were on the ground long enough to slap a coat of aluminum dope over all and give them a more civilian appearance. The Breguet's wings were set with negative stagger and slightly swept back, while the aircraft itself was easily distinguished among other war surplus types by the "rhinoceros horn" exhaust stack which projected upward from the engine. The mail was carried in torpedo-like containers set below the lower wings. It was a friendly cow of an airplane with its aerodynamic efficiency about on a par with its American contemporaries, the Jennie and the DH-4. Unfortunately, its power plant, a 12-cylinder, 300-horsepower Renault, was of dubious reliability and was a beast to arouse by hand on a chill desert morning.

American pilots flying mail during the same era knew their share of forced landings, but at least if they could walk away from their aircraft they were reasonably sure the locals would be friendly. The French were harassed not only aloft but alow. Once the safety of Rabat or Casablanca dropped behind their tail and they took up a southerly course toward Rio de Oro in the Spanish Sahara and eventually Senegal, only a few havens remained to them. These were isolated forts offering only the most primitive amenities. Landing areas were marked by empty fuel drums painted white, and the whole installation was sometimes surrounded with barbed wire to discourage visiting tribesmen.

If they were forced down between such refuges pilots could easily starve or perish of thirst before they were found, and if they were discovered by any of the constantly roving tribesmen their fate depended entirely upon the mood of the moment or the whim of the leader. In the Spanish Sahara it was said the Spanish themselves had allowed the Germans to rile the tribes against the French, for the Germans were also trying for a route through to South America and the vast Sahara stood at all compass points along the way. Even if such accusations were not altogether true there were no welcome rugs spread before the Moorish tents for either French or Spanish. Tempers were still as hot as the merciless sun when Abd el Krim, the nominal leader of the desert people, finally surrendered to overwhelming concentrations of French and Spanish troops. Even their official capitulation failed to blind the tribesmen to those fat catches flying overhead, and they continued to do their utmost to compensate themselves against such intrusions upon their age-old privacy. The very vital, swashbuckling men of the desert have never been noted for their patience, and if they troubled to demand a ransom they expected immediate payment.

Because of these unique hazards two aircraft flying in company often flew the mail through to Senegal. Thus, if one went down the other could presumably effect a rescue even if salvation for the pilot meant clinging to a wing. The Potez 25 A, a plywood and fabric biplane, was sometimes used for such escort service over the Mauritanian desert and was regularly employed by the French Air Corps which for a period flew the mails along a route farther to the east, from Algeria across the very heart of the Sahara. Later the Potez 25 As were used on Aéropostale's South American division.

As an added safety factor, native Moorish interpreters were carried on most southbound flights out of Dakar or Casablanca, and on the return legs over the Sahara pilots approved of the extra weight. They knew how even a slight misunderstanding in the desert wilderness might permanently terminate their flying.

Ten years after the First World War the parade of unnerving international events is much alleviated for the French by an incomparable vintage year. Although the grim results of the American stock market crash are already being recognized everywhere, the wines of 1929 are declared most noble and are duly celebrated. In Paris, Picasso, Gertrude Stein, and Hemingway clink glasses in solemn devotion while in America a new Swede known as Garbo is a sensation. While Australians Harry Ulm and Kingsford Smith make a new flying record between Sydney and London, a kindly Quaker engineer, Herbert Hoover, is trying desperately to keep his known world from falling down around his nation's ears.

Like every other man beginning an airline, Monsieur Latécoère has had his financial problems with the enterprise he founded so boldly. While the French government did condescend to give him a token subsidy it proved not nearly enough to carry "the line" through its period of development and expansion. Somehow there always seems to be a practiced financier waiting off stage in such situations—in this case, Monsieur Bouilloux-Lafont, an international entrepreneur of considerable reputation. In taking over "the line" he has contrived to change the name from Latécoère to the Compagnie Générale Aéropostale. The Argentine subsidiary is designated Aeroposta Argentina.

To the pilots and mechanics and radio operators, and to Daurat himself who remains in direct command, "the line" is still "the line" no matter what must be painted on the airplanes. The original agreement specified that the Latécoère Company should carry 25 percent of the airmail bound from Argentina to Europe, but actual across-the-water service will not be inaugurated until 1930.

Meanwhile, the veterans Mermoz, Vachet, and Guillaumet, plus a handful of French pilots, have been relieved of their African duties and assigned to the new routes out of Buenos Aires. Daurat himself remained in Argentina long enough to see things established and later, rather inexplicably, appointed none other than Antoine de Saint-Exupéry as the Jefe de Trafico, a post equivalent to operations manager.

SAHARA STATION, "FIVE CANS"
Potez 25a and Breguet 14a2 in Saint-Exupéry country

The South American challenge is totally different from anything these men had known in Africa. The "Haboobs" of the desert and the four summer months of heavy rains over Senegal are problems of the past. Here, if there are any natives at all they may be depended upon as reasonably friendly, but the weather on the southerly route to Patagonia is abominable. Winds of 60 to 100 knots are not at all uncommon during the winter season and there are times when the Potezes take off vertically from a standing start.

The dashing Almonacid serves as an enthusiastic liaison between his government and the French. With Saint-Exupéry he is determined "the line" will not only operate to Comodoro Rivadavia, 900 miles to the south of Buenos Aires, but all the way to the Horn itself. Considering the frequent low ceilings over Patagonia and the furious winds, the project often seems utterly impossible.

The route to the west across the Andes to Chile presents an entirely different curtain of challenges. There is still the wearying wrestling match of flying the Potez in extremely rough air, but instead of skating along a hundred meters above the relatively flat terrain, no assault upon the Andes may be safely attempted below 18,000 feet—only slightly below the Potez's absolute ceiling under ideal conditions.

Flight solely by reference to instruments is still unknown to the French, and Guillaumet is the first to lose a battle against the red-iron mountains. Caught in a blizzard he dare not lose sight of the forbidding terrain, yet he is pressed lower and lower betwixt cloud and mountain until finally every path of escape is obliterated. His body is numb with cold and his thinking groggy from lack of oxygen.*

Because of the snow Guillaumet's forward visibility is almost zero as he circles hopelessly in the only open area he has been able to find. At last, directly below, he is convinced there is the shore of a lake. Flat enough for a landing? Now with his fuel supply diminishing, the normally adroit Guillaumet is transformed into that devil's slave—a pilot with but one choice.

---

* Present-day concern over the dangers of anoxia when flying above 10,000 feet without oxygen are justified and supported by medical fact. However, until the 1940s the Andes were regularly flown at altitudes as high as 20,000 feet without oxygen for the crews. Similarly, 18,000 feet was a common crossing altitude through the passes, with cruising at 16,000 and landings at 9,000 to 13,500 feet.

He executes a reasonably smooth landing, but in his euphoria he has forgotten the treachery of drifting snow and at the very last moment the Potez flops tail over nose and halts inverted. Guillaumet is unhurt, but a lonely prisoner of the most barren mountains in the world.

Perhaps if Guillaumet had known of his coming ordeal he might have lain down and surrendered. But he is no normal man, and although Frenchmen often seem to have more than their share of human frailties no one can ever fault their individual courage.

Once he has disentangled himself from his inverted Potez, Guillaumet finds the swirling white world around him ferocious. The wind is of such force it twice knocks him down. Retreating to the Potez he contrives a crude shelter of mail sacks and waits out the tempest for two days and nights. At last the wind eases, the sun breaks through a sour milk sky, and Guillaumet emerges to challenge the Andes—a little man with barely enough sustenance for a single day, his bones already freezing in the twenty below temperature, a lonely man no larger than a molecule against the gigantic facade of the Andes.

And yet, however numbed his thoughts, Guillaumet is still a reasoning realist. Now with clearing weather some sort of search will be launched by his comrades. But the Andes are still partly masked in cloud and the search area is enormous. Worse, the Potez is painted aluminum, nearly impossible to see against the snow. And the smoke of his flares is white. In the melting spring perhaps the most determined searchers may stumble upon the remnants of his aircraft. Would the spring winds still spin the upturned landing wheels while they try to unbend the frozen corpse of Guillaumet, one time pilot of "the line?"

Then realizing he must be his own rescuer, Guillaumet commences the long walk out and down toward the Argentine plain. He is woefully equipped for a venture which would challenge any mountain climber; no ice axe, no rope, and an apparently never-ending series of vertical chasms between the site of his landing and safety.

It is a long way down and out of the snows, with every foot of the way excruciatingly difficult. Guillaumet who has gambled on himself, Guillaumet the puny human, who is soon to lose nearly all senses except for survival, lowers his head and starts walking toward the east.

He walks for five days and four nights, a frozen zombie who may pause momentarily to relieve his exhaustion but dares not sleep lest he sleep forever.

At last, near the foot of the brooding mountains, the assembly of bone and gristle that was once the man Guillaumet, his two legs and arms the best evidence that he was ever a man at all, stumbles into the arms of a peasant woman. From her hut the word of his deliverance soon reaches his comrades, and yet another ordeal in the history of "the line" is closed.

Although Monsieur Bouilloux-Lafont took over the financial control of the Latécoère Company in 1927, the full effects of his machinations did not become apparent until much later. In many ways his financial gymnastics were similar to those practiced by some American airline tycoons and even contemporary politico-industrial combinations. Bouilloux-Lafont, however, added certain spices to the mess which would give it a more typical French flavor.

By 1931 Aéropostale, still bearing the image of "the line," had amassed considerable property and hardware. It had put 42 aerodromes into operation, owned 8 ships for South Atlantic liaison service to seaplanes, staffed 70 radio stations, and had an inventory of 200 aircraft. The company had also issued 195 million francs worth of stock and now, at the very threshold of Latécoère's original dream, found it impossible to meet its enormous debts. The result was bankruptcy and liquidation, although in a last minute panic the Chamber of Deputies authorized some 6 million francs to keep a few propellers turning and at least some mail in the air. The result was predictable as the familiar pattern of government interference and suspicion by association descended upon what was once a splendid organization. Recriminations flew from Bouilloux-Lafont to politicians and from the politicians back in the face of Bouilloux-Lafont. Monsieur Flandin of the government was accused by leftist groups of having sold surplus military hardware at low prices to Latécoère who, because of the liquidation, had sold it back to the government at a profit. Even Leon Blum got into the act with virulent editorials in the *Populaire,* and incredibly, Daurat himself was accused of sabotage. The left-wing press, which had ever despised Daurat, was delighted to report that he had been seen burning the mail instead of delivering it. The ridiculous accusation was based upon the testimony of a radio operator and two disgruntled pilots whom Daurat had once dismissed. Behind such nonsense was the fact that Daurat's very devotion to detail had once compelled him to experiment with the effects of fire and water on whatever flew in his planes. Hoping to discover some way to frustrate the habits of the Moorish tribesmen who treated the international mails with scorn, he used dummy envelopes and mail sacks to conduct his experiments. He wanted to thwart the tribesmen who would slit the bags for burnooses and then have at the envelopes for money. When the fun was over they would set fire to the debris. And Daurat knew that sooner or later at least one cargo of mail would have to endure a ditching in the sea. He endeavored to find some way to preserve the mail even if it was subjected to fire and water.

Some measure of the chaos that had befallen Aéropostale and of the power of political jealousy when highly charged can be realized from the fact that in spite of protests by the large majority of his pilots, including Mermoz, Guillaumet, and Saint-Exupéry, the man who had done so much to create "the line" was summarily fired.

Although the subsequent debacle sank into a morass of lawsuits and scandalous accusations, the more fantastic of which even involved General Weygand (of World War II fame) and the French Secret Service, "the line" did continue to fly, albeit with little remaining of its former efficiency or zest.

An interim company known as SCELA conducted operations until the French government got around to the creation of something called Air France which, in the arrogant manner of all national airlines, swallowed everything including some once magnificent individualists. Soon an entire era which had sparkled with the colorful adventures of many great Frenchmen became only a memory.

They were the bold young men in boots and leather who flew the Breguets and Potezes, aircraft which were inherently a compromise as employed, since their original purpose was for the military. Approximately 1,800 Potez 25 As were built. Equipped with 450-horsepower Lorraine-Dietrich water-cooled engines, they were much superior to the relatively slow and antiquated Breguet 14s.

An astonishing 8,000 Breguet 14s in various versions were built before production ceased in 1926. Wars, bureaucratic indifference, and the withering winds of time and expediency have caused this host of aircraft to vanish except for a single survivor which is preserved in the French Musée de l'Air.

The human survivors are happily more numerous and may still be viewed in all their dignity at the Club Vielles Tiges. They have only to close their eyes to once again hear the winds of the Argentine pampas or see the red-fezzed Senegalese sentries with their long bayonets glistening in the desert sun.

Chapter Twelve

The Elegant Wings of Empire

*T*he late 1800s brought a traumatic time to mariners, for all those hearties so long accustomed to the glories and miseries of sail saw their life style change exceedingly with the advent of steam power. A few stubborn sailors kept their canvas flying until the very end, but the majority accepted the inevitable doom of themselves and their vessels and swallowed the anchor with such grace as they could manage. Those who were realists adapted their ways to steam and thereby survived.

Perhaps no other aircraft in the flying world offers a better parallel to that era of fading romance than the "Scipio" class flying boat, which displayed the British flag with considerable haughtiness along a sector of Imperial Airways route to the East. In true, empirical style it was a "pukka sahib" of aircraft and quite as anachronistic. Properly dressed passengers carried their cork helmets handy for the landings in Egypt, and they were *all* "properly" dressed.

"One must keep one's status clear along the way to Injia. . . ."

Although such quaint national pride may now be long gone with the winds of colonialism and guilt about the white man's burden, it was *de rigueur* for Scipio passengers.

While the British Empire crumbled piece by piece, the drill was to look the other way. It was not happening. Much of this conservative myopia seeped into the operational philosophy of Imperial Airways. Thus, while the Americans were already flying the Boeing 247 and creating the Douglas DC-2, the English were still setting sail in such Victorian geese as the Scipios. And sadly, when one by one they departed the scene, a way of life for aerial sailors was gone forever. So also passed one of England's several chances to dominate the skies and aircraft manufacture as it had once ruled shipbuilding and the seas.

Although commonly known as Scipios, the official designation of these candidates for the world's most rococo aircraft was "Short S-17, Kent." The English have always had the decency and taste to name their aircraft types rather than number them in the monotonous American fashion, and so the "S-17" was ignored along with "Short," the manufacturer's name. Somewhere even the "Kent" became lost, and the type was stuck with the name of the first to fly. Thus Scipio.

The year is 1932. Only three Kents are built, Scipio, Sylvanus, and Satyrus. While their number seems minute in comparison with present quantities it is a direct reflection of the aerial times. Only thirty-two pilots are regularly employed in the whole of British air transport, a number exactly matched by the number of airline transport aircraft. Across the channel, 135 French pilots are flying 269 aircraft, while the Germans somehow manage to fly 177 aircraft with only 160 pilots and still far exceed any rivals in weekly air mileage.

Apparently the production of only three aircraft has some magic quality about it which applied to flying boats. Built and flying during almost the same period (1931–1934) are Pan American's only three Sikorsky S-40s, a four-engined flying boat hoisting nearly as many aeronautical square sails as the Scipios. Also contemporary are the grandiose DO-Xs which so very clearly demonstrate that even the usually progressive Germans have not been able to put aerodynamics and flying boats together. Only three DO-Xs are ever to be built, which was just as well. Still in the future will be the last of the big four-engined flying boats, the Vought-Sikorsky 44s. Again only three will be built.

All the while Glenn Martin is brooding in Ohio and Maryland. A most peculiar individual, somewhat of a charlatan and somewhat of a genius, he has always been a stout champion of flying boats. With his eventual production of the Martin "Clippers" plus his very genuine energy and influence he will do much to keep the aeronautical anachronisms flying long past their legitimate time.

The Scipios ply the seas and sky mainly between Brindisi in Italy and Alexandria in Egypt. In so doing they are playing a vital role in aviation history for their very assignment is the direct result of a national philosophy which if carried to equal extreme by additional governments might chain international flights to the ground and cause incredible difficulties for worldwide air transport.

Soaking complacently in a tub of bureaucratic arrogance the French and the Italians have proclaimed sovereignty of the atmosphere above their respective real estate and have forbade any "over-flight" with passengers. Thus anyone bound for England to the Far East is obliged to go by sea or take a flight from London to Paris, transfer to a train, and proceed overland to the port of Brindisi. There, having paid due tribute to the French and Italian railroads, the passenger may transfer to a Scipio and resume eastward progress. The resulting shambles if every nation in the world should take such bullheaded attitudes could conceivably affect flights into outer space, a fancy which such men as H. G. Wells, Jules Verne, and Robert Goddard have visualized.

Eventually the restrictions are to be abandoned, but not without a gesture from a volatile Italian who will express his continued objection by setting fire to and destroying the flying boat Sylvanus in Brindisi harbor. Rumor will say he is a railroad employee.

Like other vehicles of transport all large flying boats were born of a common denominator—an international lack of airports. In the time of Scipios it is ever so much easier to put down a few mooring buoys and hire the local beach boys and their "bum-boats" for connection with the shore than it is to build runways. Men of almost every nationality are familiar with the sea, but very few know anything whatsoever about creating airports. The general conception of a ship that flies, especially if the route is over water, is relatively easy to accept and it will prevail until a still unforeseen world war proves the "safety" factor of flying boats alight-

ON HIS MAJESTY'S SERVICE
Short S-17 Kent (known as Scipios) taking off from the Nile at Khartoum

ing in anything but protected harbors is hardly much better than a "ditched" landplane. No one will quite believe that the occasions for unscheduled landings in the briny will be so rare that the chance may be virtually ignored in aircraft design.

While amphibians suffer the worst reputations for expensive maintenance, flying boats have always run a very close second, and both types are cursed with innumerable problems which are not present in less glamorous aircraft. Corrosion is only one persistent battle involved in flying boats, and operation in their liquid element creates an almost unresolvable conflict between aerodynamic efficiency of design and water-born demands. At normal landing speeds water becomes nearly as hard as concrete, and so the hull of a flying boat must be built over-strong, which of course takes a penalty from the useful weight load. Likewise the engines have to be fixed high above any possible spray since mere water droplets can very quickly destroy the efficiency of whirling propeller blades.

While comptrollers are frowning over maintenance costs against revenue earned, the flight operations people of any flying boat transport company are constantly reminded of those hazards peculiar to their craft, some of which just cannot be anticipated even with the best techniques. "Glassy water" landings can be damned dangerous if not accomplished just right, and *all* night landings offer more than enough opportunities for accident. Striking a single dead-head log on landing or takeoff can sink a flying boat, and a flat calm day can cause an embarrassing delay in departure while the frustrated pilot chases his tail around trying to create enough wake to break hull suction. Otherwise intelligent surface mariners have always shown a strange indifference to understanding vessels in flight, and crowded harbors can present nightmare effects to a flying boat skipper who is trying to set down between a determined ferry, an odd freighter or so, and various small craft whose helmsmen have apparently been struck blind. And then there is always the over-anxious launchman bringing either officials or supplies alongside who refuses to recognize that the relatively fragile topsides of a flying boat will not repulse his clumsiness with the same ease as a steel ship of thirty thousand tons. As for schedules, realistic flying boat operatives soon acquire a wistful air. Try as they might to eliminate mechanical delays the vagaries of wind and weather have a much greater effect on their aircraft than on landplanes. Even arrivals and departures of the reliable Scipios are occasionally two or three days late.

Yet many amenities soothe those passengers who are in enough haste to travel by flying boat. The Scipios carry only sixteen passengers which allows people a chance to stretch their legs and be treated as honored guests aboard rather than as so many faceless bodies weighing (with baggage) a uniform 170 pounds. The cuisine is excellent and served on tables covered with lace doilies meticulously squared about a vase of fresh flowers. Spotless napkins are folded in the water glasses to be removed and snapped with gastronomical solemnity by the anticipating diner. And the finest wines are served in cut-crystal goblets—the reds uncorked and allowed to "breathe" a suitable time before tasting and the whites chilled yet not frozen into a flavorless pseudo-sherbet. Nor is the passengers' peace of mind or conversation constantly shattered by inescapable pronouncements via a public address system. If anyone really cares they simply glance at the airspeed and altitude indicators conveniently placed on a cabin bulkhead. Airspeed 105 miles per hour. Altitude 1,000 feet. The windows are easily available to everyone aboard, and the view from a decent altitude only contributes to an undeniable yet strangely pleasant sense of being somewhat superior to the rest of mankind.

"Ladies and gentlemen, if you will step this way the launch will take you ashore."

The Scipios provide an elegance, a style, and a drama of self-respect never to be played again. And sometimes if a crocodile chooses to snooze on the mooring buoy urging his departure is also taken for granted. God is in his heaven and Britons, Britons . . . never shall be slaves.

All three Scipios were eventually destroyed. The namesake crashed on landing at Mirabella in Crete, and in 1938, nearly three years after the passionate Italian scuttled Sylvanus, the Satyrus was scrapped.

Two landplane versions of the type were built and dubbed Scylla and Syrinx. They flew the route between London, Paris, and Brussels and were not at all popular with either pilots or maintenance crews. Everything was out of convenient reach for mechanics, and a session at the ponderous controls left the strongest pilots physically weary.

Chapter Thirteen

Iron Annie

*I*t was so that for a time in North America and later in Central America a certain ungainly-looking flying machine became queen of the continental skies. On first sight it seemed that all the Aunt Mabels in the land had turned their washboards over to the Ford Motor Company who had decided to try a little riveting here and there and fly the result. The final production was known as the "Tin Goose" or, more formally, the tri-motored Ford, and the exploits of the type are legend.

During the same era a far more ubiquitous flying machine began its distinguished career and was soon flying in regions where the Ford was totally unknown. It was the Junkers 52/3m which eventually became known as "Iron Annie."

Mounted German Uhlans are still carrying lances and dying for fatherland and Kaiser as the year 1915 promises certain victory over the Allies. General Erich von Ludendorff, pressing on toward Paris, is the genius behind German maneuvers along the western front, and a wretched Austrian private named Hitler is bucking for promotion to corporal. Germany is not worried even though the war is not proceeding exactly according to plan. It is common knowledge that the Russians to the east are lice-riven animals and the French are decadent. There are not enough English, military or otherwise, to matter. And so victory is inevitable even though an overzealous *Unterseeboot* commander has sunk a ship named the *Lusitania* and brought down a barrage of international scorn upon the German *Kultur.*

This year a certain Professor Hugo Junkers, formerly a successful heating engineer, has become intrigued with the possibilities of aircraft for advantage in the war and later for the coming peace when Germany rules the Western world. Some say he is a little crazy because he has formed a company to manufacture an *all metal* flying machine. Metal *fly*? Obviously Herr Professor Junkers has slipped into early senility at age fifty-six.

It is hardly surprising that no one will donate precious aluminum for such a mad enterprise. Look at the lightness of properly built aircraft like the Eindeckers. Both critics and friends of the aging Professor Junkers strongly recommend he take counsel with that young Dutchman Tony Fokker before venturing further into aeronautics.

Yet a first aircraft is something like a first child, and Hugo Junkers has no intention of sharing parentage. He proceeds to build his creation alone, using sheet iron for skin and iron tubing for frame. The wing is of cantilever design, an innovation which only magnifies the madness of the scheme, and fears are openly expressed for the life of young Lieutenant von Mallinckrodt who is to fly the silly thing.

On a gray December day, when German children are already being warned that because of the war Christmas must be most frugally observed, von Mallinckrodt becomes airborne in the world's first all metal aircraft. Designated the J-1, it is the sire of innumerable Junkers aircraft, one of the most notable being the Junkers 52/3m.

Soon after the successful flight of his first aircraft, Junkers actually does join with the enterprising young Dutchman to form a new company, Junkers-Fokker-Werke AG. It survives until 1919 when the Versailles Treaty terminates all German military aircraft production. Before the boom falls, Professor Junkers's transport version of the Junkers 10 manages to make the first commercial flight of an all metal aircraft between Weimar and Dessau. The load manifest lists one passenger.

Only a few years pass before Germany is back in the air with Professor

Junkers leading the way. He is working on a flying wing design and has actually produced the Junkers F-13 with the now standard Junkers construction of corrugated duralumin skin and metal throughout. With two crewmen the F-13 is something of a pacemaker in itself. Not only does it find favor in Europe, but six are purchased by, of all institutions, the U.S. post office which intends to use them on the mail service between New York, Chicago, and Omaha.

Several F-13s are sent off to China where the German-controlled Eurasia Aviation Corporation is solid proof of teutonic interest in the future of Far East aviation. To exploit that interest two Junkers 24s fly from Berlin to Peking via Siberia and return without incident. The aircraft can carry nine passengers and is a great success on the European circuit. The year is 1920, and Germany is on the rise again.

Two years later a Junkers christened *Bremen* is employed in the first east-west transatlantic nonstop flight between Baldonnel in Ireland and Greenly Island off Labrador.

Enter ex-corporal Adolf Hitler (Iron Cross), whose views on the future of Germany and particularly German aviation do not agree with those of Professor Hugo Junkers any more than they do with the concepts of a world-renowned lighter-than-air man, another Hugo—last name Eckener. While Eckener manages to hang on albeit grudgingly, Junkers is forcibly "retired with honor," but not before his firm has produced the famous three-engined Junkers 52.

It is not long before the 52s are flying all over the world and for good reason. Their performance cannot be duplicated by any other contemporary aircraft, operation is economical, and their overall construction is so hell-for-stout they are ideally suited to the very rough going found in the boondocks of Asia, Africa, and South America.

In China, where China National Aviation Corporation operates what is possibly the most colorful flight service in the world, the Junkers 52 is more comfortable and far more efficient than the American-built Stinsons and Loening amphibians which are also on the CNAC roster. You can stuff more than twenty Chinese in the cabin, or heave up a satisfying load of ammunition for an anxious warlord, or remove his body from the field if he is no longer functional.

CNAC pilots are a recalcitrant, swashbuckling lot of internationals who make their own flight rules and promptly proceed to break every one of them. They drink when they are thirsty and are not too particular about how much time elapses before they make the next ascension. The likes of two Americans, Sloniger and Caperton, who have come to China with the intention of selling Curtiss-Wright products are quite at home among as rambunctious a covey of aerial adventurers as has ever been assembled under one corporate flag. Even Pan American, which with its usual eye to the future has contrived to buy the American interest in CNAC, cannot tame their new and merry band of pilots.

All self-respecting CNAC pilots do a little personal smuggling, but then smuggling is a way of life throughout China. They are thoroughly familiar with the rules of "squeeze" when applied to their own benefit, and their attitude toward their official employers, their Chinese hosts, and their passengers is haughty crossed with good-natured condescension.*

On one matter only are all CNAC pilots in fervent agreement. For their money, which is not inconsiderable, you cannot build a better airplane than a Junkers 52. Everything about it is necessary, and unless a man dives it straight into the ground, walking away from a forced landing is almost a certainty. Among other attributes they are pleased with the way the 52s get off the ground so handily despite the occasional overloads and the nearly perpetual steaming heat in the south of China.

Yet beautiful women and fine aircraft must always hold certain tricks to make them interesting, and as so often happens after first infatuation, a man may discover the fault is the direct result of a quality, a contradiction he missed on first appraisal.

One of the reasons for the 52's agreeable flying nature is the employment of the "double wing" concept, that is, full-span ailerons and slotted flaps. The separation between the two sets of airfoils is considerable, so much that when viewed from below the Junkers 52 appears to have a second wing trailing along behind the main span. The immediate result is great aerodynamic efficiency in the ailerons and lateral control is excellent.

---

* This reputation was hardly diminished by various of their exploits while flying "the Hump" during World War II in DC-2s and DC-3s.

IRON ANNIE
Junkers JU-52-3M at a landing field in China

Ernst Zindel, the original designer of the single engine 52 which developed into the tri-engined version, was not a pilot, and he was as ignorant as his contemporaries of a nuisance called "ice." Alas, European pilots flying 52s soon learned the hard way that their normally benign bird could become an absolute beast after only a few minutes in ice. The separation between the wing trailing edge and the aileron leading edge provided an ideal trap for ice accumulation, and there was double jeopardy in the result. If the pilot was foresighted enough to keep wigwagging his wings and thus avoid a complete lock, there was nothing he could do about the form in which the ice accumulated. Unfortunately it often produced a negative lift effect and a *complete reversal* of control response.

Deutsche Lufthansa customarily named their 52s after famous German airmen of World War I. There was the *Manfred von Richthofen,* the *Oswald Boelcke,* and the *Ernst Udet.* As tragedy struck time and again new 52s were christened with the names of Lufthansa pilots whose final plea for a DF bearing had been followed only by permanent silence. The villain was invariably ice.

By 1928 Professor Junkers's enterprise had produced more than a thousand aircraft. There was no such matching production anywhere in the world, and the feat was continued right into World War II when at peak operation the Junkers complex employed 140,000 persons. Junkers 52s had been heavily involved in the Grand Chaco war and the Spanish civil war. In all, 4,832 "Iron Annies" were built, and although thousands were lost in World War II they refused to become extinct or even moribund when that conflict ended.

Air France and even the French air force used the 52s, which were built in France during the occupation. British airways used 52s to help reentry into the airline business after the war. The 52s were similarly valued in the Scandinavian countries, in Portugal, Greece, Poland, South Africa, and in South America.

Forty years after the "Iron Annies" first took the sky the Spanish air force still keeps a few operational. There are stories of "Iron Annies" still flying for a living in New Guinea and in South America. Even if they are ghost stories their repetition is but further proof of the 52s' iron durability. Only America's own DC-3s come anywhere near such a record.

Chapter Fourteen

The Tin Goose

*H*enry Ford, never one to be overly tolerant of errors committed within the shade of his corporate nest, decided in 1926 that one of his more recent associates should be put out to pasture. It was then, as well as now, a common corporate maneuver in lieu of a blunt firing. Thus is the dignity of higher-echelon executive ranks preserved, charges of corporate cruelty thwarted, and if the victim is an executive with a long-term contract his continuous exile from such jolly strongholds as the executive dining room and his total quarantine from corporate affairs soon causes all but the most broken men to depart of their own will. Which is the object of the game. As a result, the employment contract is voided, and company funds which would otherwise have been siphoned off as salary for the disgraced chieftain are used for more promising activities.

Ford beheaded one William Stout in this manner. His sin was the creation under Ford auspices of a three-engined aircraft known as the 3 A-T, a design which possibly had more things wrong with it than any other contemporary flying machine. It was so awkward in performance, impractical, and ugly in appearance it seems almost impossible that it could have been the brainchild of such a gifted designer as William Stout. Ford considered he had reason to be disappointed in a man whom he had financed and lifted from obscurity. With characteristic high-handedness he saw to it that a convenient fire destroyed the offending aircraft together with all the design notes and papers related to the project. The first three-engined aircraft produced under the Ford banner could therefore be declared as never having existed.

The next version, the 4 A-T, otherwise known as the Ford trimotor, soon made an incomparable place for itself in aviation and has also been the subject of more legend, admiration, and misconceptions than any other single aircraft.

For various reasons William Stout, who originally conceived a high-wing all metal aircraft powered by three engines, had very little to do with the successful version. Once Stout had been removed from the pitcher's mound, Ford turned the project over to William Mayo, his chief engineer, Harold Hicks, a racing boat designer, Otto Koppen, John Lee, and James McDonnell, who would one day command the McDonnell Aircraft Corporation. Thus if the trimotored Ford had no other claim to fame it was one of the very few successful anythings ever designed by a committee.

It was a committee utterly dedicated to a single purpose—creating an aircraft that would be good enough to please Henry Ford and thereby save their collective necks. To that desire they gave all of their talents and took aid where and in any way they could find it.

One clandestine effort was typical of their desperation. The first contemporary multi-motor aircraft was the Fokker F-VII, a three-engined, high-wing monoplane which had set many admirable long-distance records. By chance, Admiral Byrd, who had chosen a Fokker for his arctic expeditions, dropped in to Dearborn where his welcome was most gracious, and his Fokker hangared for the night.

It was a long night for the committee who spent most of it bending copper tubing into templates as they measured the shape of the Fokker's wing every foot of its length. Only a very few people were ever told of the night-long measuring party, certainly not Henry Ford. But those who did know were not surprised to discover a remarkable similarity between the metal airfoil of the Ford 4 A-T's wing and a certain aircraft of Dutch design.

Although Stout was out of the vortex almost before he had a fair chance to prove himself he was by no means defeated. Nor would he accept executive banishment into the flat hinterlands surrounding Dearborn. A sort of combination Norman Bel Geddes and Buckminster Fuller, his thoughts were so progressive that one of his earlier designs, a monoplane fighter for the U.S. Navy, could well have been at home in the jet age. He was a pleasant man, a confirmed devotee of bow ties. His bushy hair complimented by a Chaplinesque brush of mustache made an imposing trimming for his wide-spaced challenging eyes. He was also a modest man who would insist that his chief contribution to aviation was interesting Henry Ford in its potentialities. Perhaps so but no honest record of air transport would deny that it was "Bill" Stout, not a committee, or the Great Henry himself, who first conceived of such a radical departure from tried and proven craft of wood, wire, and fabric. It took a man of very strong convictions to champion such a totally new concept in flight design, and fortunately for aviation and for Ford, that same stubbornness kept Stout in the corporate fold long after his embarrassing mistake.

The Ford 4 A-T carried twelve passengers and was an immediate success with the public even before very many travelers had sampled its spartan comforts and become aware of its performance. Its very appearance, the absence of flying wires along its cantilevered wing, the suggestion of massive strength in the thickness of the wing, and the sense of reassurance in its all metal construction had a powerful psychological effect upon passengers. Once the Fords took to the air, transports of conventional wood and fabric construction were doomed, and the lead position was definitely cinched when the later-model Ford 5 A-Ts with the more powerful Pratt and Whitney 400-horsepower engines were put into service.

The three engines on Fords were activated by inertial starters. On signal, a mechanic or the copilot inserted a crank in each engine separately and commenced his labors. He wound slowly at first and then faster as his efforts spun a heavy flywheel which was geared to the crankshaft. Once the flywheel was whining at what sounded like full speed, the cranker pulled a cable which engaged a spring-loaded clutch thereby transmitting the energy of the spinning wheel to turn the engine. The squealing sound was almost identical to that displeasure emitted by a resentful pig kicked in anger. If the oil was not too cold the propeller would turn at least three or four revolutions. Usually this was enough to start the engine, but if the pilot was not alert and missed the moment of truth with mixture and throttle, the crankee was obliged to start his labors all over again. Winding the cranks was very hard work and at times when the engines proved unaccountably balky there were impolite exchanges of opinion between cockpit and cranker.

The success of the trimotored Ford proved once again the powerful influence of various and often unrelated factors upon the launching of any enterprise. Once again the magic of timing, which as usual no one could foresee, turned what was inherently a good thing into a triumph. The first tri-motored Ford came off the assembly line in 1927. That same year young Charles A. Lindbergh, who had nothing whatever to do with Henry Ford or Fords, flew the Atlantic and the world went aviation mad. Everyone wanted into the act, and there was Ford more than ready with the ultimate answer to their desires.

The "timing" was now tightly wound, and as always its released energies created other situations to enrich the whole.

Among the several U.S. airlines, three fledgling outfits, each more than willing to try anything "to make a buck," began to take a serious interest in long-range passenger service. These were Transcontinental Air Transport; Western Air Express, which already had mail contracts and occasionally carried a few people who liked to be different; and Maddux Airlines, named after its proprietor, a West Coast automobile dealer who had fastened on to a very good thing—the exclusive sales agency for Henry Ford's new trimotored aircraft.

Meanwhile the newly promoted Colonel Lindbergh was no man to bask perpetually in public adoration. Once he had proved his point, which in retrospect seems mainly to have been that man is indomitable, he wanted honest work. Transcontinental Air Transport recognized his very real talents as well as his fame and hired him as chairman of their Technical Committee. The title was more than high sounding and was not intended as an honorarium. T.A.T. wanted Lindbergh to survey and literally create from a vacuum an air service which would allow a passenger to travel on one ticket coast to coast with as much of the journey as possible being made by airplane. The idea in itself was novel since the myopic railroads still did not offer a through coast-to-coast service. Rail passengers were not only obliged to change trains in Chicago but also stations, and the fast "Limiteds" took three days and three nights.

Lindbergh went into his new job with the same dedication he had given to his flying career since the beginning. New airports had to be built and Lindbergh insisted they meet his strict requirements for utility in all weathers and safety. T.A.T. had a nucleus of thirty-four mail pilots, but more would be required to fly the new service. The pilot roster included such very special personalities as Ben O. (Benny) Howard of racing fame, Otis Bryan, Harry Campbell, Ted Howe, Ed Bellande, and Jack Fry who would one day become boss of the whole airline. Sharing the pilot selection task with Lindbergh was Paul Collins, a World War I pilot, and John Collings, who as a barnstormer had been one of the first to fly the Ford trimotors.

While Lindbergh and his associates busied themselves with the technical air aspects of the blooming T.A.T., the powerful railroads continued to cooperate in what would ultimately seem to be their own obsolescence. Whether they realized what they had wrought will never be known, but certainly the timing, the influence of Lindbergh's dignified charm, and the immediate availability of an impressive flying machine did much to press their judgment.

At least everyone understood that no sane passenger would endure a continuous flight all the way across the United States, and so something had to be done to keep the bodies moving east or west while they slept. Otherwise whatever time might be gained by flying would be lost in spending at least two nights in midcontinent hotels.

As a consequence the air-rail scheme came into being. Passengers in New York would board the "Airway Limited" and roll along through the night as far as Columbus, Ohio. In the morning a special air bus designed by none other than the ubiquitous William Stout would transfer the passengers from the train to a trimotored Ford and it was up and away to St. Louis and Wichita.

Finally, the Ford would descend through the evening light to Waynoka, Oklahoma, where once again the passengers would entrain. Dining and Pullman car amenities were furnished by the Atcheson, Topeka, and Santa Fe, which rolled on through the second night to Clovis, New Mexico. There, on the windswept plains at a station appropriately named "Portair," the passengers once again enplaned for the balance of their flight to Los Angeles. They landed at Winslow, Albuquerque, and Kingman en route. When things worked as they should it was at last possible to cross the United States in forty-eight hours.

In these same United States it has long been axiomatic that any endeavor worth the attention of a publicity flack must be inaugurated by a quota of celebrities who need not necessarily be aware of exactly what they are sponsoring. There must also be buttons to push for dignitaries who might otherwise strew obstacles in the path of progress. All of these ingredients were abundant on that July day in 1929 when the air-rail service made its first departures. At New York's Pennsylvania Station, Amelia Earhart, who had been appointed "assistant to the General Traffic Manager" of T.A.T., christened a Ford which was on static exhibition, and Dorothy Stone, dancing star of Broadway musicals, christened the observation car of the "Airway Limited" which responded by chuffing away on schedule with the first load of passengers. Since these were the times of Prohibition the easy access to French champagnes through anyone's bootlegger was ignored and all of the christenings across the land were committed in grape juice. Perhaps these unseemly acts were responsible for the air-rail service lasting hardly more than a year.

In Los Angeles the ceremonies were heavy with movie celebrities. While Bell and Howell and Universal newsreel cameras were hand-cranked, both Mary Pickford and Gloria Swanson were invited to christen one of the two Fords on the line, Douglas Fairbanks smiled his multimillion-dollar approval, and the profundities of civic officials were made available to the masses via radio. Finally, none other than "Lucky Lindy," who had the night before pressed a button in his Los Angeles office to signal the "Airway Limited" out of Pennsylvania station, now taxied the first Ford to the end of the runway and took off for the east. Joy was unbounded in the T.A.T. publicity offices. But those who labored there were far from finished for the day.

In spite of a gray and drizzling morning a considerable crowd attended the ceremonies at Columbus, Ohio—takeoff site for two westbound T.A.T. Fords. This time it was Secretary of Commerce Lamont in Washington who had his turn at button pushing. His forefinger activated a gong at Columbus Airport which in turn sent the T.A.T. pilots roaring off to earning their flight pay. After both planes had disappeared toward the west, an inauguration breakfast was held for the assorted civic and state officials who had considered their presence on such an occasion imperative. As a part of their reward they were privileged to exchange solemnities with tycoons Henry Ford, his son Edsel, and Harvey Firestone. Speculations on the success of the air-rail service were dutifully optimistic. Only a spoilsport would correctly predict that in the first month of operation a mere 153 passengers would be carried coast to coast.

One of the factors which had originally inspired the air-rail service was the lack of suitable airway lighting along T.A.T.'s route. The Fords were confined to "daylight-only" operations—which in the beginning was perhaps just as well. For pilots soon discovered certain troublesome and potentially dangerous flaws in this otherwise splendid flying machine. While the Ford engineers had diligently copied the Fokker's attributes they apparently went back to their native Model T thinking in cockpit windshield design. The glass was set vertically which created a most confusing display of reflections from any lights on the ground. Until pilot protests brought about a change, night approaches and landings on the outskirts of cities were spiced with illusion. Then a further weakness was discovered in the wing just aft the cockpit. In their pursuit of weight reduction the engineers had chosen to cover the wings center section with a very light gauge of corrugated metal. The area developed a nasty habit of cracking under the Ford's ordinary vibration, then peeling backward in flight to set up a baffling effect on the slipstream which passed over the tail. The immediate and violent shuddering caused several pairs of sweating pilots to whisper the name of their maker in the air and express profound gratitude for their physical strength when they finally reached solid ground.

For a time the restrictions on Ford night flying created some curious social and economic conditions among pilot personnel. Since it was unlikely that any of the U.S. airlines could long survive without mail payments the mail was transferred to single-engine aircraft at night and continued on its way while the Fords slept. As a result two different categories of pilots were employed. One group flew the mail at night just as they always had—armed with helmet, goggles, leather flying suit, an ability to fly solely by instruments in all but the most severe weather, plus an almost uncanny "pilotage" ability. Another group flew the multiengined Fords and Fokkers and led a more comfortable but considerably less exciting life. The separation of job categories dissolved once the passenger ships were approved for night and instrument flight. Then a ladder became available which enabled a copilot on the passenger flights to eventually become his own master in a single-engined mail plane while mail pilots moved on to become captains of the larger aircraft.

Once all the button pushing was done and the broken glass of christenings swept away, the U.S. post office sat back to take a hard look at something for which it had been largely responsible. And Postmaster General Walter F. Brown was not altogether pleased. By 1930 the McNary-Watres Act had provided him with the clout he needed, and Brown was not in the least shy about using it. Among his new powers were the sweetmeats of "route certificates" which amounted to an exclusive right to fly at a profit over most parts of the country. It was furthermore Brown's privilege to award the various routes to those airlines he considered most capable. His criteria was simple and honest enough—performance, financial stability, and potential for expansion.

"Competitive bidding in the airmail business," he announced, "is of doubtful value and more or less of a myth."

This forthright if somewhat undemocratic approach was the beginning of the end for many small airlines, which was exactly what Brown wanted. He liked big, strong, well-knit entities flying his mails and to encourage them he was not at all hesitant about so arranging the franchises that connecting airlines would be forced to join bodies as well as extremities. Thus for months there was a tremendous flurry of airline executives and financiers fighting both for survival and the spoils.

One of the subsequent mergers involved T.A.T. which married both Western Air Express and Maddux Air Lines in a ceremony as devoid of sentiment as the connubial arrangements for three Ubangi families. Combining the three names was too much of a corporate mouthful and appeared unsightly even when printed on a Ford's ample flanks. The handle was abbreviated to Transcontinental and Western Air, and TWA* was born.

Soon after the honeymoon the new triumvirate abandoned the air-rail show and settled down to pure flying business. One of the spurs toward this new efficiency was the embarrassing record of N.A.T. (National Air Transport) which was flying the northern transcontinental route straight through in Fords and advertised only thirty-one hours westbound. They also advertised the eastbound trip as only twenty-eight hours.† Also since N.A.T. was in the process of becoming United Air Lines the spanking new TWA realized they would have to hustle, a total effect which canny Postmaster Brown had desired all along. Even so, TWA had to refrain from night flying passengers until the route lighting was completed. Until then transcontinental passengers spent the night in a Kansas City hotel.

The crash of a trimotored Fokker with famous football coach Knute Rockne aboard marked the demise of the Fords' only serious rival. Investigators declared the Fokker had suffered a structural failure because of its wooden construction and the CAA demanded so many modifications their continued operation became impractical. Ironically then, the very aircraft which owed so much to Tony Fokker's design, became undisputed queen of the skies.

Aerial royalty or not, life with and aboard the trimotored Fords was far from ideal. For American passengers there was little comparison between the cold box lunches tossed into their laps by the copilot, and the soup-to-nuts-on-a-white-tablecloth cuisine offered by equivalent British and European aircraft.

Ford passenger cabins were always too hot or too cold and the decibel level assured them a top place among the world's noisiest aircraft. Immediately on boarding, passengers were offered chewing gum which would allegedly ease the pressure changes on their eardrums during climb and descent, but it was just as much to encourage a cud-chewing state of nerves. They were also offered cotton which wise passengers stuffed in their

* Later Trans-World Airlines.
† The considerable effect of prevailing westerly winds upon such a slow aircraft as the Fords is all too apparent in the difference.

TIN GOOSE
Ford Trimotor in a mid-west thunder storm

ears so they would be able to hear ordinary conversation once they were again on the ground. Wise pilots employed cotton for the same purpose, and many who scoffed at such precaution had trouble meeting the hearing requirements on their physicals for years after their service in Fords was done.

In the first Fords there were no seat belts. Only hand grips were provided to stabilize passengers, and summertime flying could become a purgatory. While they bounced around in low-altitude turbulence the passengers muttered about "air pockets" and a high percentage became airsick. Even with a few windows open the cabin atmosphere developed a sourness which only time and scrubbing could remove.

If the passengers retreated to the lavatory they found little comfort at any season. In winter the expedition became a trial by-refrigeration since the toilet consisted of an ordinary seat with cover. Once the cover was raised for whatever purpose there was revealed a bombardier's direct view of the passing landscape several thousand feet below, and the chill factor in the compartment instantly discouraged any loitering.

The long transcontinental trips were best endured by confirmed masochists, although over the western part of the United States there was often one redeeming feature. It was a rare occasion when all a Ford's seats were occupied, and a word to a sympathetic pilot or copilot could alleviate matters. Frequent passenger Will Rogers was among the first to discover how the backs of the newer-type Ford seats could be lowered flat enough to meet the one just behind and so create a passable bunk. In addition to their mandatory list of maps, flashlights, and mail pistols, many pilots carried a small wrench to make the seat adjustment.

For the flight crew a day's work in a Ford required resourcefulness and considerable muscle. Advice from ground stations were often minimal thanks to unreliable radio gear which was located in the rear of the cabin. Frustrated pilots often applied the swift kick cure to the aggravating black boxes, a procedure much frowned upon by technicians who nevertheless had difficulty explaining why it sometimes produced results.

Even in smooth air flying a Ford became a chore if only because it was so difficult to keep in trim. The man who could coax a Ford into flying hands-off for even a few minutes was temporarily in luck and probably did not have any passengers. Even a normal bank in a Ford was an experiment in muscular coordination mixed with a practiced eye for anticipation since whatever physical input was directed to the controls a relatively long time passed before anything happened. To stop or reverse as desired *any* maneuver required a keen sense of anticipatory delay. In rough air these delays and willfulness were compounded, and just keeping the Ford straight and level became a workout. In a thunderstorm or lone squall the pilots sometimes wondered who was in charge of affairs.

In the devious way of legends the Ford has somehow emerged as a stable aircraft. In the sense that they were always controllable and therefore safe the recognition is true, but no more is deserved. Like most aircraft the advertised cruising speed was exaggerated: 105 miles per hour was much closer to honesty than the promised 120. Legend also has it that Fords could be landed in any small cow pasture, which was not so, particularly if loaded. The ability lies somewhere in between, depending on many factors, including the hunger status of the pilot or airline. Takeoff performance was actually more remarkable than landing speeds, which were much higher than most historians seem inclined to acknowledge. Perhaps they do not realize that behind the more spectacular short-field stops made by Fords which actually touched down between 65 and 70 miles per hour, there was a grunting copilot pulling for all he was worth on the long bar which extended from the floor upward between the two cockpit seats. The bar activated the hydraulic brakes which in their own obstinate way gave airline mechanics perpetual trouble.

Like any star of stage or screen the trimotored Fords bred gossip. One particularly uncomplimentary canard concerned the inability of two engines to do more than stretch the Ford's glide if one should fail. This was nonsense. At least a few pilots made a practice of shutting down the center engine to conserve fuel if the head winds were strong and a landing for refueling was inconvenient or meant even greater delay in ground time.

In 1934 TWA retired their Fords from active passenger service and in 1936 transformed a few into aerial freighters. The operation was mainly conducted out of a cinder-patch airport in New York City, known as "North Beach," a site used for the present LaGuardia.

The amount of air freight business available was still a matter of conjecture, as TWA soon discovered. U.S. conditions were quite different from those in Honduras where a New Zealand swashbuckler named Lowell Yerex used Fords to successfully exploit the distance between jungle and seaports. Within a few weeks TWA's freight operation expired from its own lack of weight. Yet elsewhere in the world, and particularly in Mexico, the Fords hauled every conceivable type of cargo with one devoted exclusively to hauling coal.

The Fords refused to die—for several very sound reasons. While the original Fords cost $48,000 fully equipped, and with tanks filled courtesy the Ford Motor Company, they could be bought for $10,000 or less once the airlines started phasing them out. Moreover, 198 were built before production ceased with relatively few lost in crashes or by other attrition. The surplus brought all kinds of aerial adventurers to the surface. They realized that here was an aircraft which was still in its prime and with the exception of the brakes could be maintained easily with a minimum of skills.

The surplus Fords were a bonanza for barnstormers who had fallen on parlous days as their two- and three-holer biplanes became increasingly difficult to fly at a living wage. Flown by barnstormers the Fords hit the familiar summer circuit carrying uncounted thousands over small towns and county fairs, fourteen passengers at a time slurping on ice-cream cones and waving out the open windows. Harold Johnson, and earlier Bob Hoover, created an air show routine which spoke eloquently of the Ford's inherent strength if a man had the strength to match it. He not only made loops directly from takeoff, but with more altitude he also did loops with the center engine shut down. He even managed a ponderous barrel roll and a languid spin.

It has been nearly half a century since the first Fords came out of Dearborn, yet some are still flying. American Airlines has two beautifully restored which are used on public relations tours. In Reno, gaming mogul Bill Harrah preserves one. Island Airlines still flies three out of Port Clinton, Ohio, and Johnson Flying Service in Missoula used Fords in forestry work until very recently. There is one in the Henry Ford Museum, and at least three more in reasonably sound health, the prized possessions of devoted antiquers.

Before William Stout died at the age of seventy-six, there was very nearly a full-scale revival of the airplane which at least originated on his drawing board. Convinced there was a place in the world's outbacks for a slow, high-performance freighter, a California group gathered enough energy and money to build something called a Bushmaster 2000. Emerging from retirement, Stout himself worked on the redesign, which from a distance seemed a haunting reproduction of its Detroit ancestor. One was actually built and flown, but soon after Stout's passing, the enterprise vanished into that mysterious limbo occupied by so many other well-intended aeronautical adventures.

Chapter Fifteen

The Dutch East India Men

*A*mong the ruling nobility of aircraft families certainly the Douglas clan has long displayed an honored crest. Yet like all families the dignity and sometimes even the honor of the name may be threatened by a rapscallion whose character cannot be easily explained. Who knows who really sneaked in the castle postern door when the true conception was arranged?

While the official version of the Douglas DC-2's birth remains the pompous pap which airline publicity offices and aircraft manufacturers are so inclined to issue, those released contained subtle hints that all was not normal during the creation of the aircraft which soon established itself in the hearts of various formidable men.

The mischief began in 1932 when the redoubtable Jack Frye of TWA told the Douglas company his airline would be a willing customer for ten trimotor transport aircraft. Obviously he envisioned something more modern and efficient than the current trimotored Fords.

Enter Colonel Charles Lindbergh who, in his role as Chief Technical Advisor for TWA, asked for an extremely sticky guarantee. If the Douglas company wanted the business, the proposed aircraft must be able to climb to a safe altitude after failure of one engine on takeoff from any point on the TWA route.

The result for TWA was not unlike that of a man who sets out to buy a balloon and comes home with a helicopter. Instead of a high-winged tri-motored aircraft Douglas midwifed a twin-engined low-winged affair which they christened the DC-1. It bore almost no resemblance to any of its Douglas forebears, nor for that matter to any other native aircraft. It was suspected that through his friendship with Donald Douglas even Antony Fokker bore a hand in its design.

The powers at TWA soon recovered from their shock thanks to the DC-1's remarkable performance, which included a successful single engine takeoff from Winslow, Arizona (elevation 4,938 feet). There were changes to be made but the DC-2's sponsors knew the sweet smell of being ahead of the game. The nearest competitor was the Boeing 247, still the stalwart of United Air Lines.

After specifying certain structural changes an order was placed for twenty-five aircraft. Perhaps this was when the actual bastard was conceived. Or possibly the DC-1, of which only one was ever built, was a hermaphrodite.

The eventual result was such a different flying machine it was designated the DC-2; and lo, the fathering of a whole new generation of aircraft was done. And simultaneously innumerable lies, deceptions, and slanders, some of which survive, were widely circulated. TWA advertised the DC-2 as a 200-mile-per-hour airplane, which it was not by some twenty miles per hour. Pilots assigned to fly the DC-2s were advised to take a Charles Atlas course in muscle building. Unnecessary. It was the *copilot* who furnished the beef by hand pumping the flaps up and down and sometimes the landing gear. On demand he set himself to the task by "activating" a long metal bar beside his seat known as a "Johnson bar." Origin of the name remains obscure, but comparison to "Swedish steam" from sailing ship days seems most likely since the total energy output matched a long spell at an anchor windlass. In summertime the more fastidious copilots carried a small towel to wipe the sweat from their brows, thus hoping they would offend neither their captains nor the single stewardess.

The Captain of a DC-2 also developed a mighty left arm although his exertions were of relatively short duration. When taxiing the braking system in a DC-2 was activated by heaving on a horn-shaped handle protruding from the left side of the instrument panel. By simultaneous use of rudder and handle the desired left or right brake could be applied. Since there was inevitably a lag between motion and effect the DC-2 was stubbornly determined to chase its own tail on the ground and in cross winds, sometimes switching ends to the embarrassment of all aboard. Taxiing a DC-2 was an art rather than a skill, and even chaste-mouthed pilots were occasionally given to blasphemy.

The DC-2 engraved a further quite unexplainable blot on the Douglas escutcheon. It could and often did make proud men humble. For reasons which are still obscure, consistent good landings in a DC-2 were impossible, or if claimed, a proof of that man's self-delusion. A pilot might make ten beautiful landings in a DC-2 and the eleventh, regardless of related factors, would cut him back to size. The stiff-legged landing gear was part of the trouble, but never was a pilot heard to complain of it when his landing was smooth. A few men could successfully three-point a DC-2 but not always gracefully, and each time they knew they were stretching their luck. Most pilots made wheel landings followed by an immediate and very firm forward shove on the control yoke to make sure the DC-2 stayed put, for once a DC-2 started bouncing the performance became so spectacular only Herculean efforts plus grim determination could tame the beast. After a series of such humiliations even veteran pilots were known to stand glaring at their DC-2s while morosely considering another line of work.

Two other idiosyncrasies contributed to the gristle and bone life of DC-2 pilots. When flying in rain the cockpit windshields leaked so badly the effect within was that of a seriously depth-bombed submarine. No amount of caulking seemed to improve the fault. The steam heating system, centralized about a boiler in the forward baggage compartment, was alleged to have been designed by Machiavelli. Much of the time the contraption gagged, gurgled, and regurgitated ominously, and to the dismay of the passengers occasionally filled the entire interior of the aircraft with vapors. A copilot charged with adjusting the several valves intended to control the system could make or break his esteem with both crew and passengers according to his success in keeping temperatures somewhere between intolerably hot or intolerably cold.

A further reminder that a DC-2 was not quite a gentleman's airplane came when the flight had arrived at destination and the engines were shut down. Then the copilot was obliged to crawl back into the rear baggage compartment, a small cubicle located just forward of the tail. There, while maintaining a praying position, he passed out passenger luggage, mail, and cargo to a ground crewman. If he became thoughtful in the process it was not because he knew he was erasing the press in his uniform pants. He was kneeling on *1,000 pounds* of sandbags, a permanent adjustment to the weight and balance problem in a DC-2.

Yet DC-2s had a pair of endearing qualities which easily balanced their sins. They were built hell-for-strong and pilots caught while trying to sneak through a thunderstorm were profoundly grateful for that strength. Wintertime pilots beset with ice knew that if *any* aircraft could survive the trouble they were in, the DC-2 was unquestionably the mount of their choice.

While DC-2s were capturing the bulk of flying trade on U.S. airlines they were also serving on the most exotic route in the world—the famed Indies service of *Koninklijke Luchtvaart Maatschappij*, pronounced by non-Dutchmen as KLM.

Amsterdam. A September day, 1935. Captain Wilhelm Van Veenendaal reports to KLM Operations at Schiphol Aerodrome where he signs for a briefcase containing monies of twenty-four countries and a table showing that day's convertability of pengös, rupees, marks, and drachmae to the guilder. He is a big man, hearty of voice and manner, as are so many Dutchmen. He is one of KLM's strangely international roster of pilot greats—Smirnoff, Geysendorffer, Koppen, Frijns, and Parmientier. Also on the list are two Germans, a Swedish count, one Australian, two Englishmen, and a handful of recently recruited Americans.

Life is good and even rather grand for KLM pilots. They are well paid and highly respected. Their opinions and advice are heeded by the front office people and their morale stands high. But they work hard for their honors, flying as much as one hundred and seventy hours a month.

DUTCH EAST INDIA MAN
Douglas DC-2 over rice paddies near Bandung, Java

And sometimes more. They must fly the abominable European winter weather relying as much on their cunning as on the questionable bearings offered by French and Belgian ground stations. Zero-zero landings are a common task. At least at London's Croyden Aerodrome the British demonstrate their thorough understanding of heavy fog. After an aircraft has landed the officials have the courtesy to send out a lorry with a 24-inch red searchlight. A pilot follows that welcoming red orb to the invisible terminal.

Van Veenendaal's schedule to the East Indies is grueling. He takes off in the predawn for Budapest, then on to Bucharest, thence to Athens where he brings his DC-2 to the arid Greek earth just before dark. There Van Veenendaal and his crew—copilot and radio operator—must sleep fast for they will be called for duty again at three in the morning. For the next several nights, even in their slumbers they will have trouble stilling the monotonous drumming of Wright engines which persists in their ears.

Takeoff is again at dawn with the first flight leg to Crete, then Alexandria, and Bagdad by lunch. The 120-degree heat is beginning to tell on both passengers and crew. On the ground the interior of the DC-2 becomes a cauldron. Yet life aboard a DC-2 is still infinitely more comfortable than flying Air France's DeWoitines (known as "Anteaters"), or worse, Imperial Airways' "Hannibals" which make the same flight in the broiling depths of 500 feet while the DC-2 has been cruising at 12,000.

That second dusk, descended from the cool heights, the DC-2 is brought to earth again at Basra on the Euphrates. While the passengers are taken to a hotel, Van Veenendaal and his crew elect to sleep on the roof of the terminal building. It is cooler and desert stars instead of bugs decorate the ceiling.

The third day it's on to Jask in Persia where the DC-2 slips down to a sandspit outlined by five-gallon oil cans. The terminal consists of a customs shed and a fly-infested canteen. A few mangy camels give the passengers something to photograph while Van Veenendaal's patience is tried by the second worst customs in the world.

The dubious honor of being the world's number one offensive bureaucrats is incontestably held by the officials in Karachi, the DC-2's next stop. There, KLM crews know it may take as much as three hours to satisfy the functionaries, particularly if they are trying to make a few honest guilders on the side. The kind of men who fly this route in DC-2s are individualists or they would not be involved. Almost as an instinctive reaction they have built a lively personal trade in mica from Calcutta, star sapphires and camphor from Burma, paper pengös to Budapest in exchange for silver—"little" things in need of transport which will not effect the DC-2's actual load or be awkward to carry.

At nine that night a landing is made at Jodphur and now it is a weary Van Veenendaal and crew who rest so briefly in an old palace of red sandstone. For once again they must rise before dawn and be off to Allahabad which is the worst aerodrome on the route. The heat is punishing for man and penalizing to the DC-2's performance. Allahabad, in the heart of India, presents an aerodrome only 2,800 feet long which makes for some soul-searching during a midday takeoff. Later in this long day there is Calcutta to come which is not much larger and finally, after crossing the Bay of Bengal, there will be a night landing at Rangoon on a field designed for "floaters" like the old trimotored Fokkers and Junkers—not for DC-2s.

The fourth day it is lunch in Bangkok then on to Penang with a landing on a short strip surrounded by high jungle. For most of the flight east of Athens charts have been suspect or grossly inaccurate. Misplacing a town or a river is easily forgiven, but mislabeling the altitude of a mountain by a few thousand feet creates bitterness in all airmen. So all of the way Van Veenendaal has navigated by bearings received from ground stations, landmarks he has viewed before, and an occasional Adcock radio range. Most useful tool of all has been his pilot sense, an instinct carefully nurtured over the years until now it is automatically triggered at the approach of aeronautical evil or falsity in the void below.

The Malacca Straits slip beneath the DC-2's wings all during the afternoon; then it's Singapore for the night after a sensual landing on a pleasant grass field. Here is at last some refreshment for Van Veenendaal and his crew. They enjoy a bottle of excellent Singapore beer before bed, a cooling ocean breeze during the night, and superb pineapple for breakfast. At five A.M. of the fifth day.

Then once more it is off into Kipling's thundering sun, to Palembang in Sumatra and to Batavia, which is now known as Djakarta, in Java. Kipling's sun is also setting on the Dutch colonialists who have so tightly held this lush empire from the days of the fat "East India men" sailing ships to the DC-2.

Reward is nearly at hand for Van Veenendaal and his crew. Once the passengers and mail are unloaded they take off for Bandung in the hills, a cool retreat where they can rest for seven days before taking a physical and once again tackling the long haul back to Amsterdam. Bandung is in a high valley and offers a fine grass strip. While the flight crewmen stare sleep-eyed at the bougainvillaea, excellent Javanese mechanics go over the DC-2. The overhaul is complete even though only a change of oil is mandatory, for KLM maintenance is so exacting even a half-inch separation of the throttles on takeoff is considered intolerable.

One hundred and thirty-eight DC-2s were built and flew under the flags of twenty-one countries. As the DC-1 begat the DC-2, so did the DC-2 beget the DC-3—an event which influenced the development of air transport all over the world.

Chapter Sixteen

The Masterpiece

A time would come when a great host of people would claim the honors for inspiring the Douglas Sleeper Transport (DST), an aircraft which soon emerged from its beautiful cocoon and became the legendary DC-3. As in all human affairs every masterpiece acquires a multitude of latter-day sponsors, but in the conception of the DST two men stand in a fort of solid truth. They are the remarkable and much beloved C. R. Smith, who then headed American Airlines, and William Littlewood, the line's chief engineer. An extraordinarily enterprising pair, the pithy, venturesome sort of men who originally forged the airlines of the United States, they persuaded Donald Douglas to expand his already successful DC-2s into a "sleeper" version. The eventual result was an aircraft which resembled the DC-2 only when viewed from a great distance.

If the DST was justly rated as the queen of all contemporary aircraft then her aerodynamically identical sister, the DC-3, must be considered the simple working member of the family, short on glamour while blessed with productivity and hence assured of a very much longer life. As in fairy tales it was the humble chargirl who became famous and exerted a powerful effect upon world history.

The DC-3 was and is unique, for no other flying machine has been a part of the international scene and action so many years, cruised every sky known to mankind, been so ubiquitous, admired, cherished, glamorized, known the touch of so many different pilot nationals, and sparked so many maudlin tributes. It was without question the most all-up successful aircraft ever built and even in this jet age it seems likely the surviving Douglas DC-3s, full-blood sisters to the long forgotten DSTs, may fly about their business forever. One DC-3 flown by North Central Airlines logged 84,000 hours air time as of June 1966, more than any other aircraft in the history of flight.

A late October afternoon in 1936. Newark Airport in New Jersey, the only airline terminus for the entire metropolitan area of New York. The world is uneasy—everywhere. Hitler has reoccupied the Rhineland in defiance of the Locarno Pact. The new and very bloody civil war which erupted in July has already torn Spain to pieces and the defunct League of Nations has abandoned Ethiopia to the Italians. Yet for Americans life is not at all bad. Rib roast sells for only 31 cents a pound, and track star Jesse Owens has won four gold medals in the Berlin Olympic games. Among other amenities, well-funded people can now avoid the interminable rail journey from coast to coast by booking passage in one of American Airlines' glistening DSTs.

The operation is now routine, which means that the much-publicized "Mercury" flight to Los Angeles is supposed to depart Newark at 5:10 and touch down on Glendale airport at 8:50 tomorrow morning—God and the elements willing. The average has been reasonably good, but the prevailing westerly winds across the North American continent often make the public relations people wish the schedule was printed on elastic.

For the fourteen passengers there are certain consolations to curb their impatience and the two movie celebrities who have paid more than the $150 standard fare for the privilege of occupying the private compartment known as the "Sky Room," are duly and regularly soothed by the not-so-subtle attentions of the stewardess. Once in the air, all souls aboard with the exception of crew will be served cocktails on the house followed by a fine steak dinner. Then Captain Dodson will send back his written flying report to be passed among his guests. It will be signed by the copilot, who will be pleased to answer any questions during one of his tours through the cabin.

One hour after takeoff the DST drums sonorously westward in the last of the autumn twilight. Eight thousand feet below the land is enveloped in full night and only the electric fires of civilization maintain the reality of motion. Memphis is still six hours' flight time over the murky horizon.

Captain Dodson is an old-timer—helmet and goggles airmail. He wears two stripes on his sleeve, ostensibly of gold braid in the maritime style, but actually the material is a cheap imitation long since faded until it is only yellow ribbon. Dodson, who is balding, wears his uniform cap continuously when on duty. He clamps his headphones *over* his cap, enduring the entire awkward and heavy combination because like so many other veterans of open cockpit flying he feels undressed in an aircraft without something on his head. His copilot, still a nameless and faceless nonperson as is the lot of all copilots, identified by one and one-half pseudo-gold stripes, keeps his left headphone away from that ear lest his master offer priceless comments on the desirability of the stewardess and her possible willingness once they are returned to earth, or the effect of rumored aircraft additions on the copilot's seniority, or perhaps, some verbal jewel on the beauty of the night. And like all copilots sensitive to the fashion of their mentors he wears his cap throughout the flight.

"We have more cross-wind than forecast. Figure me the wind."

The copilot checks his flight log for past position times and headings. He takes a round computer from his shirt pocket and turns it to the reverse side. Holding his small flashlight in his teeth he makes a pencil dot on the drift scale and twists the clear plastic disk until the dot lies across the true course line.

"I get ten degrees at twenty-five knots."

"Ah? We have a little help . . . maybe."

But doubting, Dodson looks down at the abyss below and meditates on the lights of a town sliding slowly beneath the left engine.

"No. I think we are in the middle of things. The wind is right on our beam."

And since Dodson is the captain that is where the wind is.

Later both Dodson and his copilot will be obliged to depart separately from the cockpit and make their way along the darkened aisle of the passenger cabin. It is common knowledge among airline personnel that regardless of uniform you can always differentiate an American Airlines pilot from those who fly for United. United equips their aircraft with Pratt and Whitney engines which are marvelously smooth in operation. Through corporate expediency American relies on Wright engines, faithful enough but cursed with such inherent vibrations the kidneys of all American pilots are sorely tried. Their frequent trips "to talk with the passengers" invariably terminate in the lavatory.

In spite of their unromantic mission aft "Mercury" pilots experience a curious emotion once they close the cockpit door. They may walk with a trifle less assurance, allow even a momentary hesitation in their silent parade. For there is no more stunning reminder of a man's aerial responsibility than to pass along the line of berths, the green curtains closed and swaying ever so slightly in response to the motion of the aircraft. Behind those curtains are individuals, trusting and utterly dependent on two total strangers who for this little time spell must directly control their destiny. The pilots are only ordinary men who must now relieve themselves.

If such brooding might create an atmosphere of melancholy in the cockpit, Dodson and his copilot can easily dismiss it. They may take refuge in minimizing their roles in the fast-developing scheme of the flying business.

"Full load again? I'll tell my relatives the company must be giving free rides to the coast."

"The Pullman cars must be hurting, but I guess it'll be ten years before I make captain."

The copilot's guess is deliberately pessimistic. If he keeps out of trouble there is every chance he will be wearing two full stripes after less than two years with the line.

As American Airlines pilots, both Dodson and his copilot fly for the "Flagship Fleet," which affects certain trappings designed to impress customers and please news photographers. Each DST is named after a state of the union, and a flag bearing the double eagle insignia of American is hoisted above the right cockpit window. The "colors" ceremony is carried out by the copilot who opens his window immediately after the landing roll is completed and inserts the flagstaff in a special socket. Huzzah! But copilots are very human and subject to many distractions even if they are dignified with the title of "First Officer." And sometimes they forget to douse colors before takeoff, creating an unmistakable coolness in the attitude of the captain. In his adamant opinion a man who is paid $190 a month for the privilege of sitting on his right should not forget—*anything*. Captain Dodson, who flies sleepers because the pay is higher than daytime schedules, earns as much as $9,000 a year and if *he* forgets something only God, his wife, or his chief pilot dare accuse him.

It is a clear night and Dodson will follow the long line of winking airway lights into Memphis. Having made a wheel landing in the new style rather than a 3-pointer which might awaken his passengers, he will sigh and sign the logbook with his name and seniority number.

Now, yawning mightily, Dodson pulls his flight bag from the baggage compartment just aft the copilot's seat and proceeds to the cabin. If he forgets to remove his name plate from the cabin bulkhead, the copilot or the stewardess will do it for him and moving quietly down the aisle they will descend into the soft Tennessee night one after the other. They are met for briefing by the new crew who will take the "Mercury" through to Dallas. And even outside the aircraft their voices remain subdued.

"The gyro precesses. I wrote it up."

"The man in number six snores like a walrus. The woman in upper three says wake her so she can see the dawn. Number eight is a please-do-not-disturb."

Memphis to Dallas and thence to Phoenix where the woman in upper three becomes ecstatic over the newborn desert day. And once aloft again yet another stewardess shakes each green curtain decorously.

"Breakfast will be served in forty minutes—"

Two hours later, but twenty minutes over scheduled arrival time, fourteen reasonably contented passengers disembark at Los Angeles-Glendale airport. They are refreshed by their long night's sleep and ready for the new day—which is considerably more than can be said for the irritable time-lagged passengers who have endured a flight over the same geography in a present-day jet.

The Douglas sleeper (DST) was comfort with wings, and it was possi-

NIGHT BIRD
Douglas DST Sleeper over Nashville, Tennessee

ble to treat every passenger as a very important person. As a consequence the appeal of the long Pullman journey across the continent began to fade and thus did air transport move into a new and vitally important phase. Very soon after the inauguration of the DST flights the sister DC-3s were in the air plying their trade so efficiently no established line could afford to be without at least a few. Eight hundred and three commercial-type DC-3s were eventually built, and over 10,000 military versions.

The DSTs (née DC-3s) are an easy aircraft to fly, almost totally forgiving to the most ham-handed pilots. Their inherent stability makes them an excellent instrument aircraft and their low stall speed combined with practically full control response at slow approach speeds allows the use of very short fields. They can be slipped with full flaps or held nose high and allowed to descend in a near power stall. In the hands of a skilled pilot DC-3s can be successfully landed in just about any cabbage patch some optimist has dared to call an air field.

Actual DC-3 performance records have caused frequent head shaking among those who could hardly believe what they knew was fact. Civil type DC-3s were licensed to gross 25,500 pounds. When World War II came along both ground and flying people became ever more casual with weights stuffed in airplanes. A gross of 31,500 pounds was not at all uncommon for DC-3s (C-47 in the most prolific military version), and no one bothered with such trimmings as weight and balance forms. Except for the placement of spare engines, which were a highly concentrated weight per cubic foot, the rule was to "shove everything as far forward as it will go." Once a China National DC-3 designed like all others to carry twenty-one passengers carried *seventy-five* passengers out of China to a Burmese air field. Among them was a certain James H. Doolittle who had commenced his journey on the aircraft carrier *Hornet* and was thence routed via Tokyo to an unscheduled landing on the Chinese mainland.

More than one thousand DC-3s (C-47s) took to the air on D-Day in 1944 and they were among those present at all the major World War II engagements. They kept right on flying throughout the Berlin airlift and continued through the Korean conflict.

DC-3s have been put to almost every conceivable aviation use. They have been used as makeshift bombers, glider tows, forest fire bombers, ambulance litter ships operating directly out of battle zones, and equipped with rapid-fire guns they have even served as fighters in Vietnam. Most airline pilots referred to them as "Threes," the Air Force dubbed them "Gooney Birds," and the Navy, "R-4Ds." The British chose "Dakotas," and the Russians, who were presented with a great number, conveniently forgot where they were built and were soon manufacturing their own version.

Then there was the famous "DC-2½" a crippled CNAC aircraft, which flew to its home base quite sedately with a DC-3 wing on one side and a DC-2 wing on the other. There were more than fifty different military versions descended from the original DST.

Regardless of the label the Douglas masterpiece was profoundly admired by all who flew it.

And still is. Wherever you go in the world there is a very good chance you will see a DC-3 at work, and in some of the less sophisticated regions there is a very strong possibility you will be in one.

Chapter Seventeen

The Clippers

Captain H. Lanier Turner touched thoughtfully at the beginning stubble of beard on his jaw and yawned. It had been a long night. With the early morning sun at his back he became once more aware of that peculiarly stale and unwashed feeling which descends upon airmen as they approach the end of a long over-ocean flight. Half-empty coffee cups were scattered about the flight deck, and while the animation of Turner's crew had revived with the sun, they soon subsided again into long silences—waiting quietly in the limbo of every long flight's last hour.

Turner was captain of a Boeing 314, the Pan American Clipper *Anzac*, on this benign December morning. He was squinting at the cumulus clouds just beginning to build their daily monuments across Molokai strait when the radio operator handed him a written message.

"Case 7, . . . Condition A."

Far to the west, with the dawn just breaking, Captain Hamilton, commanding another Boeing 314, the Clipper *Philippine*, eased his ship off the turquoise waters of Wake Island lagoon and took up a course for Guam. He was still climbing for altitude when his radio officer handed him a hastily scribbled message.

"Case 7, . . . Condition A."

Hamilton immediately turned back for Wake and descended at full power.

The identical message was flashed to the Clipper *Pacific*, Captain Ford, then en route for Auckland from Noumea.

No one had to translate the messages for the three captains. Though they read with natural disbelief (had it finally happened?), their pre-flight briefings included information which history has shown was not at all surprising to a great many people, including the president of the United States. What President Roosevelt knew would be, and what so many key people in the Pacific air business knew, had at last transpired.

The Japanese had attacked. For weeks the only question had been exactly when and where. The least awareness sadly enough was at Pearl Harbor, whence Captain Turner and his Clipper *Anzac* were bound. According to prearranged plans he changed course for Hilo, landed for refueling, and by evening was once again airborne, this time eastbound with all haste for San Francisco and home.

Captain Hamilton, caught at Wake Island, endured a Japanese attack which occurred soon after he landed. His Clipper *Philippine* was machine-gunned from nose to tail, but miraculously remained airworthy enough to escape with all of Pan American's island-based personnel except one. That unfortunate was so occupied driving an ambulance he missed the Clipper's departure. Waldo Raugust's reward for his self-sacrifice was capture by the Japanese.

On the East Coast of the United States, Sunday afternoon was a time of relative relaxation for two powerful Americans. They were alike only in their penchant for power, and Sundays merely slowed the pace of their restless minds. One of these men was Glenn L. Martin, a tall, austere, bespectacled personage who, among other things, considered himself a superior sportsman. The judgment was his own. A lonely man, even those who knew him casually had little enthusiasm for his analytical approach to the kill. During the duck-hunting season on his Maryland estate, he concealed himself and such male companions as he could muster in an unique sort of blind. They enjoyed certain advantages over other hunters, for Martin's grain fields were wired for sound, and strategically placed loudspeakers repeated recorded mating calls. The blinds were not constructed of rushes or other natural grasses, but of large mirrors so arranged they offered the illusion of endless rows of corn stubble. Lured by the mating calls and the visual tranquillity, a great many ducks never knew what hit them.

Glenn L. Martin was also a fly fisherman, but he was not one to stand whipping his rod all day in the hope of a lucky strike. Instead he would leave his Royal Coachmen, his Blue Duns, and his rod behind while he conducted a two-day survey of the stream he intended to fish. Only after he was thoroughly acquainted with those pools and rills haunted by the local trout and their dining habits would he resolve to engage the prey. He was rarely skunked.

The sporting methods of Glenn L. Martin were only partly indicative of the man. Obsessed with an Oedipus-like devotion to his iron-willed little mother, he was also dedicated to flight and had been ever since the days of the Wright brothers and another pioneer, Glen Curtis.

Something of a seer, Glenn L. Martin had long forecast that the day of the huge flying boat would eventually arrive and lift the only business he knew out of the experimental category. On that December afternoon in 1941 the gray-haired, now almost legendary man, who had hardly known even the temporary company of any woman except his mother, brooded about the catastrophe at Pearl Harbor and the immediate need for large flying boats. Much to his chagrin the newer Pan American Boeing 314s had recently replaced his graceful Martin Clippers, which could carry only 47 passengers on short flights and a mere 18 overseas. The Martin price tag was $400,000 each, which resulted in an over-ocean per-seat cost of more than $20,000. Such economics, typical of Martin's curious indifference to other people's financial problems, were too much even for Pan American.

Although his Clippers had been disappointing, Martin had no intention of swerving from his flying boat beliefs. He very resolutely predicted the time when enormous flying ships, carrying as many as one thousand passengers, would girdle the world. He already had various versions of such enormous aircraft on the drawing boards, and he was convinced that his now vast organization would soon be engaged in their production, first for the military and later for commercial use. The factories and thousands of employees which comprised the Martin Company were the current result of his youthful enterprise—those long-gone days when, with the encouragement and help of his mother, he had built his pusher biplane in an abandoned California church.

Now, the great and wealthy man, who had literally invented barnstorming and had eventually produced multitudes of aircraft including several of the first flying boats, still consulted the power behind his throne, his mother.

That same Sunday afternoon at his Long Island home an entirely different sort of individual had little time to brood, for the demands of his empire were suddenly compounded. He was Juan Terry Trippe, sometimes known as Juan "Napoleon," whose major sport was hunting and capturing air routes throughout the world. In this dedication his methods were quite as calculated and ruthless as were Martin's with fish and fowl.

Trippe, the high cockalorum of Pan American Airways, had already established himself as an aviation baron in much the same way acquisitive railroaders Vanderbilt and Jim Hill built their empires. Trippe's deliberately kept low profile had given him a reputation for mysterious maneuverings akin to Sir Basil Zaharoff's in the arms industry. He was a pipe and cigar smoker (supposed evils Martin would never have permitted himself), and seen in his office high in Manhattan's Chrysler Building he might impress the casual visitor as "a nice Ivy League boy who's found something to do." While it was true that he had married a Stettinius, and both business and social associates included such blue bloods as the Whitneys, Vanderbilts, and Harrimans, Juan Trippe spent the greater part of his waking hours with a totally different cast of characters. These were the hard-working operational officers of Pan American: Andre Priester, the Chief Engineer; George Rihl; and none other than the ubiquitous Charles Lindbergh, whose exploratory and inspirational contributions to Pan American were considerable.

None of these men, nor any other employee of Pan American, worked any harder than Trippe himself, who seemed to thrive on international conquests. A State Department official once described him as being more pro-Pan American than pro-American. Obviously he entertained the notion that he and his airline were ordained with a divine right to do all of his nation's overseas flying. This conception, reinforced by the Kelly Foreign Air Mail Act of 1928, so subsidized the successful bidders on international routes that they were able to enjoy about the same control over trade as that once granted by King Charles II to a group of gentlemen adventurers known as the Hudson's Bay Company. Trippe with his very influential connections had a way of hearing about new opportunities long before any competitors and almost invariably got what he was after.

Trippe's methods at route acquisition were not unlike Glenn Martin's ways of the hunt. Both men were cold calculators, not chance takers. In Trippe's case that meant keeping domestic politicians perpetually off guard and thus out of his hair. If the government wanted an air route to China, Trippe's Washington spies learned of that desire long before any possible competitors. He would then dispatch his emissaries on specific missions, their oriental targets selected not by chance or intuition but by logic. Meanwhile Trippe himself frequently dropped in on friends long established in the Victorian elegance of the State Department and the considerably less colorful but far more money-minded post office department.

Every official Trippe talked to was aware that Trippe moved among the great and influential of the land, and consequently they were well disposed toward his presence. They found his quiet charm so irresistible that when bidding time came they were pleased to consider more than the mere numerals in his application. Trippe invariably bid the highest allowable dollar for a mail subsidy while depending on certain secretly held aces to secure a winning combination. On the route to China the federal officials would discover that only Juan Trippe and his PAA held exclusive landing rights to Hong Kong or Macao, a circumstance which conveniently eliminated less foresighted bidders regardless of price. Likewise did the Caribbean and most of South America become PAA's private preserve, a bounty it held for nearly two decades.

It would be very unjust to assume that Juan Trippe operated from the same self-serving platform occupied by so many corporation executives since the ancient Greeks found it advantageous to combine their business interests. Nor did Trippe at the height of his career greatly enrich himself.

In 1942, at a time when PAA truly lorded it over all other airlines with its worldwide operations, Trippe's salary was only $23,000 plus some not extravagant legal fees and a share participation in PAA's dubious stock.

Trippe was a business man, but he also thought of himself as a patriot doing his best to make his country supreme in the skies. That he vigorously opposed any other combination trying the same thing was beside the point. He argued that to the debit of his country foreign governments would certainly play one applicant for air rights off against the other if there were more than his anointed self in the field. He also maintained there was no such thing as freedom of the air, a stand which in light of most governments' jealously guarded skies was also true. In the case of the United States, whose name alone created a pathological fear of Yankee imperialism in the minds of many foreign governments, a private company could reasonably hope for much greater concessions if left to fight the overseas battles alone. The lonely warrior and that company must naturally be Juan Terry Trippe and his Pan American.

As is the way of all born leaders, Trippe's attitude and philosophy seeped down through the ranks and eventually became the prevailing atmosphere in which PAA employees conducted their operations. While Trippe himself was anything but outwardly arrogant, too many of his subordinates seemed to interpret his subtle imperialism as license to disdain the rest of the flying world and even what there was of the flying public. The consequence was a wave of passenger resentment and fury unmatched to this day.

In the 1930s passengers booked on PAA and expecting to be treated as valued guests soon had their views readjusted. They discovered Pan American was doing them a great favor by allowing them aboard any of their aircraft, and once present and accounted for they were to do precisely as PAA and particularly the captain *ordered* them to do—right johnny-now and no complaints. In the early 1940s, as the pressures of war made every seat a hard-won prize, the passenger who was not a V.I.P. was subjected to cavalier treatment unknown to other airlines. From the time the passenger bought his ticket to his final landing and release from captivity the haughtiness of many Pan American personnel was almost too flagrant to be taken seriously. That a great many people did so remains one of the few

UP AND AWAY BEFORE THE FRONT ARRIVES
Boeing B314 Clipper taking off west-bound across the Atlantic

identifiable reasons other airlines were finally permitted to trespass on Pan American's hard-won routes.

Strangely enough there was never any reason for such self-consciousness. Pan American with her flying boats unquestionably became one of the world's greatest airlines and in many ways the most outstanding.

One of Pan American's most enduring boasts (echoing along with their "world's most experienced airline") has been a Soviet-like tendency to claim "firsts" in everything. A remarkable number of those claims were true or nearly so, particularly in PAA's over-ocean pioneering. While the passenger-handling department perpetuated its sorry record, operational crews developed excellent techniques and flight planning procedures which were copied by many international airlines. "Howgozit" curves were but one example and precomputed celestial navigation from aircraft another. These tricks of the long-range trade were passed on to the military at the beginning of World War II and later proved invaluable to the rapid growth of the Air Transport Command.

Pan American was the first line to establish nautical tradition in the air, referring to the commander of their aerial vessels as "captain." In the same fashion the second pilot was referred to as the "first officer" and the mechanic as the "engineer." Carrying the theme further, time aboard prewar PAA Clippers was kept in bells and progress in knots.

Trippe liked nautical smartness, and from the very beginning of Pan American he declared there would be no leather jackets or breeches worn by his flight crews. Thus they were the first to wear a common uniform, albeit so somber and lacking in insignia even the wearers often referred to it as their pallbearer rig. Not until long after competitors had adapted the navy-type four, three, and two gold stripes indicating rank did PAA crews abandon their unadulterated black.

The original maritime orientation was an inevitable result of Pan American's enduring affair with flying boats. Their very character influenced flight operations and employer environment.

A newly hired Pan American pilot could not help but be intimidated by the long haul ahead of him. Prior to World War II, PAA standards were extremely high with no less than five degrees of advancement available through home-study courses. The idea was to provide a career rather than merely a job. Starting in the lowest grade of apprentice pilot, the young man was obliged to work in the traffic department, and maintenance, communications, meteorological, and stewards' departments, supposedly long enough to appreciate what was keeping him in the air. His pay during this period was $175 per month, and occasionally he would be assigned to a flight crew. Once aboard a Clipper, he was advised in unmistakable terms that his job was to observe, to stay out of the way, and to keep his mouth shut. He also knew that if the allowable gross weight was exceeded he would be the first item unloaded.

The next degree was junior pilot, achieved only after obtaining a license as engine and aircraft mechanic, radio operator, *and* celestial navigator. After these achievements he had to pass tests in international law, maritime law, seamanship, and at least one foreign language before he could qualify as second officer on a flying boat. Even then he was usually signed aboard as "supernumerary."

The next step upward was service as flight engineer, which led in due time to senior pilot rating. Presumably he was then qualified to command any PAA aircraft available although there was still a final degree in the hierarchy. This would be "master of ocean flying boats," a category so exalted that very few Pan Am pilots ever achieved it. The basic idea behind the rating would have warmed the hearts of all good dues-paying members of the Airline Pilots Association since a master of ocean flying boats was entirely relieved of the controls. Again in the maritime tradition he was simply in command with Pan American capitalizing on his experience while the graying pilot prolonged his career.

Other requirements and restrictions seemed ridiculously harsh to many airmen, including some who flew for PAA, yet those demands had a way of fostering pride as well as contributing to the safety record. Any pilot, regardless of his previous experience, had to have 2,000 hours of PAA flying before he could take an over-ocean command. Five round trips across the Pacific were required to qualify on the China route. And practically all PAA pilots were officers in the Naval reserve.

The thoroughness and quality of Pan American's flight training was superb, and for a long time the line had the pick of young applicants even though the pay was somewhat less than that offered by domestic operators.

All of this talent and expertise was at its prewar peak when the Boeing 314 Clippers arrived on the scene.

Seattle's mud-ugly Duwamish River flows past Boeing Field where innumerable Boeing aircraft have been born. It is May 1938, and nearly three years have passed since Pan American asked for certain very special designs from the Boeing, Douglas, Consolidated, and Sikorsky aircraft companies. They wanted a huge long-range flying boat, which in concept was a direct extension of Juan Trippe's ambitious thinking. The Caribbean is already his, and the necessary squatter's rights in the Pacific are held down firmly if expensively by Pan American's Martin 130 Clippers and remaining Sikorskys. Yet, as in shipping, the true prize is the Atlantic, the blue ribbon route where the money remains. Even at this early stage of the race Trippe finds it annoying and embarrassing that his Pan American is very much in trail position. Britain, even with a sporadic service rendered by her *Canopus* Empire Class flying boats, is currently as dominant in the Atlantic skies as her giant Cunarders are on the surface.

It is a questionable time for a United States finally emerged from the greatest of depressions, a time of political soul-searching and reform led by Franklin D. Roosevelt, who intends to pack the Supreme Court if he fails to have his way. Two of America's sweethearts are Shirley Temple and Jean Harlow, with *Gone With the Wind* breaking all box office records. The Japanese have launched a new and prophetic invasion of Chinese territory, and all of the four Kennedy boys are alive and maturing. Their father is ambassador to the Court of St. James.

In spite of these nervous times, Trippe has committed his Pan American to the Boeing Company for their winning design—a matter of $5 million for six Boeing 314s. Things are not going well and there have been countless delays. When finally the first 314, christened the Clipper *Honolulu*, is ready for launching into the Duwamish, a team composed of R. J. Minshall, Wellwood Beall, Al Reed, and test pilot Eddie Allen are soon challenged by all the standard difficulties involved in the production of any new aircraft plus a whole bundle of new problems they had not expected. Even before maiden flight they are already wondering if Boeing and Pan American have not attempted to be *too* innovative. There are the new type engines, Wright Cyclone 14s, which burn a new "high" octane fuel graded 100. No one has used it before. The hydromatic full-feathering propellers are new in design.

Most perplexing of all are the 314's embarrassing aerodynamic faults. During one of the earliest high-speed taxi tests they almost lost the *Honolulu* when an ordinary gust of wind caused her to dig a wing tip into Lake Washington. After much debate it is decided the only fix must be an alteration in the angle of attack of the short water wings which extend outward from the lower part of the hull. More time and more money lost.

Next it is discovered there must be something very wrong with the design of the 314's empennage. Lateral stability is very unsatisfying in rough air, and Eddie Allen claims holding the big boat on a steady course is like herding a reluctant buffalo. The final fix is a pair of vertical stabilizers, one on each side of the tail.

All of these expensive affairs and frustrations are kept so in-family that the 314's reputation remains unsullied until at last, in January 1939, they are granted an approved CAA certificate. Aeronautical historians may well speculate upon the role of Pan American—and particularly the perpetuation of large flying boats—if less than a year after Eleanor Roosevelt christened the Clipper *Yankee*, a still slumbering war had not awakened in a rage and promptly despoiled every ocean.

As soon as they slid down the Boeing ramp into the Duwamish and Pan American Captain Joseph Chase had made an acceptance flight, the 314s were flown south to avoid the Washington state tax. They were soon placed in active service and ponderously yet very surely began proving Juan Trippe had been right and perhaps Glenn Martin as well.

PAA pilots found the 314s very superior to their previous charges, the Martins and Sikorskys, and PAA comptrollers occasionally discovered encouraging signs in their books.

The 314 proved a stable instrument aircraft, easy to land on water during day or night, and altogether displayed very few faults. They were difficult to taxi downwind. Pilots departing for the Pacific from the narrow channel at San Francisco's Treasure Island were often sorely tried while barely missing breakwaters, pilings, buoys, and other marine hazards.

Once in the clear, getting a fully loaded 314 "on the step" was relatively easy, but gaining enough speed to separate so much mass from the water required patience and at least two miles. In time the 314s developed cracks along their main wing spars. PAA inspectors drilled small holes at both ends of the cracks and stopped the disease before it became epidemic.

Throughout their otherwise brilliant careers the 314s' extraordinarily long wing (only 5 feet 5 inches shorter than the mighty 12-engined German DO-X) presented taxiing problems for their apprehensive skippers. An unexpected gust or shift in wind, or too tight a turn combined with a moment's clumsiness in engine use, could dip a wing in the sea. The effect was more than embarrassing since salt water was scooped into the cabin air-circulation system, and the resulting cleanup job was awesome.

Captain Joseph Chase commanded the first 314 to Honolulu. Captain Harold Gray and his crew made the survey Atlantic flight in Clipper *Yankee,* followed by Captain LePorte who flew from Port Washington to Lisbon in 26 hours and 54 minutes while simultaneously establishing the first scheduled Atlantic airmail. R. O. D. Sullivan, captain of the Clipper *Dixie,* made the first scheduled Atlantic flight with pay passengers in June 1940. While Hitler bided his time during the so-called "phoney" war, wealthy U.S. citizens and important European personages flew the Atlantic in luxury not known since the days of the big Zeppelins.

Meanwhile Pan American's Pacific operations with 314s also promised to fulfill Glenn Martin's prophecies and Juan Trippe's ambitions, even though supercaution to the extent of making a test flight before every scheduled departure was standard procedure. The Clipper *California* set the style with a round trip to New Zealand in 101 hours and 52 minutes. By July 1940 a regular mail service was scheduled fortnightly to Auckland, and a month later passengers were accepted. To Trippe's considerable satisfaction, and as if dictated by Glenn Martin himself, Clipper arrivals and departures were dutifully reported in the *Shipping News*. Except for a bad habit of burning out their engine-driven generators, troubles with the 314s remained few.

Then came "Case 7, . . . Condition A." And the western world grew up in a day.

Only twelve Boeing 314 Clippers were built. Their utilization and their eventual fates were hardly as anyone had originally envisioned.

Soon after Pearl Harbor, Secretary of War Stimson was empowered to requisition all commercial aircraft. The Clippers *American, Anzac, California,* and *Capetown* were assigned to the army, designated as C-98s. The *Yankee, Atlantic, Dixie, Honolulu,* and *Pacific* were acquired by the U.S. Navy. British Overseas Airways bought three Clippers and promptly dubbed them with proper Saxon-sounding names, *Bristol, Berwick,* and *Bangor.*

To increase payloads, all 314s were stripped of their luxury trimmings and went to work carrying everything from tank parts to radars and toilet paper. Their licensed gross weight was conveniently increased from 80,000 pounds to 88,000, with Captain Joseph Chase once lifting 90,000 off San Francisco Bay. All grace and dignity was not lost immediately however, since the 314s performed many special missions with the world's great aboard, including Roosevelt and his staff to the 1943 Casablanca conference. With the delightfully boisterous Irishman Captain Kelly Rogers of BOAC at the helm, Winston Churchill pronounced the 314 his favorite during a flight from Bermuda to the United Kingdom.

In spite of their exposure around the world only one 314 was lost during the war. The Clipper *Yankee* caught a wing tip in the Tagus on a night approach to Lisbon and crashed. Yet long before the war's end the future of large flying boats was being resolved. While the fat 314s were lumbering their way across the oceans, sleeker and much swifter landplanes were accomplishing the same thing with much less fuss and expense. At war's end, Pan American, having sold their Clippers to the military for $900,000 each, certainly did not want them back. Like the other airlines who had also eased into the over-ocean business, Pan American had discovered the wonderful Constellations and DC-4s.

As if to mark the end of an era, the Clipper *American* made a final PAA flight between Honolulu and San Francisco in 1946. There were only 34 passengers and they were in the air over 16½ hours.

Like aged elephants seeking a peaceful place to die, the Clippers disappeared one by one from such familiar haunts as San Francisco Bay and Pan American's marine base at La Guardia field. The *Honolulu* was sunk by naval gunfire after engine problems obliged her to a forced landing in the Pacific. The majority of the remaining Clippers were declared surplus by the War Assets Administration, and it was not long before the usual bargain hunters for the postwar charter airlines found their price irresistible. And quite as usual their subsequent efforts to make money flying outdated aircraft failed. The *Capetown*, re-christened *Bermuda Sky Queen* by an English charter line, landed in the Atlantic after her fuel aboard proved inadequate to strong westerlies. Like the *Honolulu*, all her crew and passengers were rescued before the Coast Guard sank her as a hazard to. navigation. Eventually all the others were scrapped.

Glenn Martin's death coincided with the funereal cycle of the Clippers, and his passing marked the end of total believers in big flying boats. There remained only certain odd exceptions. There were the four-engined "Sandringhams" and "Solents" produced by England's Short Company, which continued active service well into the 1950s. Again in England, the Saunders-Roe people produced an enormous flying boat called the "Princess" class. Only three were built before the government finally ran out of patience and money. Although they were never put into commercial service, two of the behemoths were still occupying a ramp at Cowes in the early 1960s.

The final exception, as well as the ultimate large flying boat project, was and is Howard Hughes's fantastic *Hercules*, an 8-engined creation with a 320-foot wing span. She might well have originated in Glenn Martin's dream book. With Hughes himself at the controls, she made one flight in 1947 to an approximate height of 20 feet—just to prove she could. She remains to this day guarded in a special hangar near Long Beach, California, an aeronautical Gargantua preserved indefinitely in a Hughes-style glass case.

Chapter Eighteen

The Ball-Bearing Airline

*T*ime and the time-honored pact between airline officials and pomposity have often contrived to twist or occasionally conceal the original reasons behind the inauguration of many flight services. Thus in this jet era few stockholders realize an enormous flying corporation is the ultimate result of a tacky love affair between a financier's wife and a barnstorming pilot, while another exists because of the machinations of a union leader who thought air transport was a natural for teamster expertise.

There have been some wild stimuli behind various airlines, and many were nursed along with no thought of direct profits. The dismal defeat of the fine airlines developed by U.S. steamship companies after World War II was an outstanding example of savage corporate in-fighting and government conniving to reserve American skies for a then chosen few.

Ranking as one of the most unique as well as one of the most dangerous nonprofit airlines was British Overseas Airways Corporation's "Ballbearing Airline," which operated from the United Kingdom through enemy-held territory to Sweden during the darkest days of World War II.

The year is 1943 and all Europe with the exception of Sweden and

ESCAPE
De Havilland DH-98 Mosquito over the North Sea

R.PARKS.

Switzerland is locked in Teutonic chains. Stan Musial has won the most valuable baseball player award, a horse named Count Fleet the Kentucky Derby, and St. Louis and New York are winning pennants in their respective leagues. In England the Sitwells are reading poetry for the benefit of the Free French, and in Italy Mussolini's gathering problems are complicated by his playing host to syphilis. Unbeknownst to most of his condemning countrymen a very brave civilian who simply signs the logbook "C. A. Lindbergh" is flying combat against the Japanese.

Now more than ever the necessity for moving people and things from A to B has become important to British survival, and while Churchill is already beginning to declare "the beginning of the end," the basic strength of his antagonists is still undiminished and their control of the European skies becomes the decisive factor on every flight regardless of purpose.

KLM (Koninklijke Luchtvaart Maatschappij), which had only yesterday provided a long and very special service from Holland to Batavia, is now dormant except for a small unit flying DC-3s from England to Lisbon and Gibraltar. Their charter is under BOAC, who are also displaying the very stiffest of upper lips on a route through to Sweden.

Thanks to the Swedish habit of playing both ends against the middle, the Swedes have again managed to draw the disdain of their Scandinavian peers, the Norwegians and the Danes, and there are many inhabitants of this planet who would like to see their vaunted "neutrality" labeled for what it really is— outright comfort and assistance to the Axis powers. Their neutrality allows the Swedes to sell their superb ball bearings to the highest bidder, and the Germans are keenly aware that somewhere along the line of production every instrument of war demands the use of ball bearings. Indeed, the Germans are constantly reminded of this fundamental need by the British and the Americans who are expending tremendous effort trying to destroy various German ball-bearing works. So it is that while wearing their gaudiest abhorrence-of-war masks, the Swedes have become house dealer.

Fortunately for the English, ball bearings make a handy cargo for aircraft. A great many amounting to considerable weight can be packaged in a very small area.

So the name "Ball-Bearing Airline" has become the ordinary identification for BOAC's hardly ordinary operation between Leuchars, in Scotland, across the Skagerrak to Stockholm.

There are several other important reasons for the run. Except for telegraphic cable, the air remains the only link between England and Sweden. Flight is the only means by which British diplomats can arrive on the scene in Sweden and hope to counter Nazi propaganda. They must convince the Swedes that in spite of present difficulties England not only intends to keep fighting, but proposes to win the final battles. Also much prisoner-of-war mail passes through Sweden, and occasionally British airmen shot down over German targets have managed to escape to Sweden. If they can be brought home somehow the risk is worth it.

BOAC began the service with a single Lockheed known as "Bashful Gertie" and Norwegian crews did most of the flying. Yet even though the service had been augmented by additional Lockheed Hudsons it was soon recognized that something more versatile was required. The Lockheeds are too slow and their ceiling so limited that they are easy meat for the German fighters who are beginning to take a nasty interest in the proceedings. Interception of these "airliners" is aided by a German guest of the Swedes who faithfully alerts the *Flugwachkommando* of every BOAC departure.

The only answer to this most peculiar airline problem is the twin-engined Mosquito which happens to be the world's fastest operational aircraft. Such is the priority of the ball-bearing run that BOAC maintenance personnel are soon painting their newly acquired Mosquitos with civilian identifications where roundels and squadron designations would normally appear.

One of these aircraft is a Mark VI flown by F/O Gilbert Rae. With him on a night cursed with moonlight is Radio Officer Payne and an unidentified V.I.P. passenger, presently subject to very un-V.I.P. treatment. Acutely aware that the Mosquito was not designed for the comfort of first-class airline passengers, he is crunched in a tiny, windowless, plywood-lined compartment which was formerly the bomb bay. He is so confined he has been able to make only the most minute changes in his position since departure from Stockholm. For comfort he has a thermos of coffee, a reading light, and the very occasional voice of Pilot Rae in his earphones. And perhaps his prayers, for on this dreadfully luminescent July night the Mosquito has just been jumped by a pair of German fighters.

It is the second such unnerving experience for pilot Rae since he started flying the run, and he is doing what he can to survive. He had been cruising nicely along at 23,000, congratulating himself on the fact that his Mosquito had thus far not succumbed to its unfortunate windscreen icing habit. As a consequence Rae had been able to spot a pair of entwining contrails bound from the moon in his direction and he had almost immediately proved that wooden aircraft can often take more than their crews.

Rae can only guess at his true airspeed during the interminable twisting dive from 23,000 to sea level. High-speed stall has the needle wriggling meaninglessly against the pin, but photo-recon Mosquitos similarly equipped have been known to record 392 miles per hour at 22,000 feet while slamming full supercharger to the Merlin engines.

Certainly the speed in this desperate dive for the surface of the North Sea has exceeded 400, but how much more Rae will never know. He is too busy for such details. His evasive action has been so violent Radio Officer Payne has taken a beating and will require a fortnight to recover from his injuries.

With Rae's very own British stiff upper lip trembling slightly from prolonged fright, he has reason to wonder why it is that with several hundred Allied bombers bound for the heart of Germany at this very moment, a single unarmed aircraft bound *away* from the continent should be chosen for pursuit. And he could kiss the red-hot lips of the screaming Merlins on this particular aircraft, which is not equipped with the usual "saxophone" exhaust shrouds.

The multiple ejector exhausts make him a better target at night, but because of their propulsion effect he can squeeze an extra ten miles an hour out of the Mosquito. Rae is also grateful some camouflage expert did not paint this particular Mosquito lamp black. That paint job alone would have cost him an additional (and almost incredible) 26 miles per hour!

Seconds can so easily become an eternity.

With those handicaps in his race for life Rae is certain the FW-190 pilots somewhere in the sky behind him would long ago have cried the code word "kettledrums" to their controllers and their cannons would now be blasting him out of the sky.

This is airline flying?

Suddenly the Mosquito is alone and Rae knows it as surely as the moon reveals the sea so close beneath him. The Germans have abandoned the chase as hopeless even for the swift FW-190s, and Rae is free to concentrate on the shortest course for Scotland—and perhaps the enigmas of mercy.

Since they are civilians, both Rae and Payne are awarded the OBE and the MBE instead of DFCs. Rae will not be able to contemplate his for long since exactly one year later, while again bound home from Sweden, he will disappear forever.

Considering the circumstances, the BOAC ball-bearing operation had remarkably few losses in 4 aircraft, 8 crew, and 2 passengers. As a by-product of the route certain admirable exploits contributed to the established British reputation for audacity and indifference to hardship. In January 1944, while some airline pilots were squawking about the quality of the coffee aboard their comfortable airliners, Captain John Henry White of BOAC flew a Mosquito on the ball-bearing route from Stockholm to Scotland, then back to Stockholm, and once again back to Leuchars, Scotland, for a total of 9 hours and 36 minutes flying time and only 45 minutes on the ground. He made three crossings through enemy skies, all at night, all on instruments, and *all by hand flying*. If White had been an American it is conceivable that the Airline Pilots Association would have had a word to say about such exploitation of the working man.

The Mosquito itself was a revolutionary concept in military aircraft, and at its first introduction it met a chilly welcome from members of the RAF hierarchy, who often displayed even more obtuse thinking than our own star-bespangled types. A wooden aircraft? Preposterous! Unarmed? Absolute blather, old boy. Only two chaps in the crew? Who flies the bloody thing if the pilot is killed?

Geoffrey de Havilland had his problems with military fustiness and bureaucracy, but when he finally won out he had an aircraft that would carry as large a bomb load to Berlin as a B-17. If necessary his Mosquito could go it alone while the B-17 risked an eleven-man crew and an expensive four-engined airplane, plus whatever fighter escort was required.

The "ball-bearing" route was finally flown with such determination that a contemporary traffic problem developed with as many as a dozen aircraft slipping through the Skagerrak at the same time. A considerable part of its success was the Mosquito's ability to "get the hell out of there," and the astounding record the Mosquitos accomplished in Europe alone makes them the aerial escape artists of all time.

In January 1944 a Mosquito flown by Warrant Officer Kennedy was jumped by *nine* FW-190s. A full-power dive and escape was made northwest of Beauvais after which Kennedy calmly photographed Paris, Chartres, and Chateaudun. Then he went home for tea.

Later in 1945 two Messerschmidt 163 rocket fighters caught Flight Officer Hayes at 30,000 feet. He half-rolled and dived to 12,000 feet passing through an IAS of 480 miles per hour. Hayes got away to be bonged with a DFC and managed to convince all concerned that the combination of iron men and wooden ships was not quite obsolete.

Then just after D-day invasion, as if yearning for love as well as admiration, Mosquito fighter bombers employed their drop tanks in the speedy delivery of beer to the beachheads. It was unofficially estimated that the gesture resulted in more military progress per sortie than the original intent of the Mosquito design.

While it is true the Mosquito can only be considered as a freak among air transports, it forever proved its worth en route to priceless ball bearings.

Chapter Nineteen

The Purebred

*M*ore than half a century has passed since the scratch beginnings of transport aviation. While the dash and peculiar character of the men involved changed very little until the advent of the jet era, the aircraft they flew were internationally a very mixed bag ranging from clumsy and dangerous peasants to splendid aristocrats. When the jet transports at last ruled the skies everywhere the considerable differences between aircraft greatly diminished. Regardless of their conformation those jets remaining in production beyond the prototype models are safe, easy to fly, and most of them are economically efficient.

The combination of these virtues has washed much of the color from the flying scene, and the overall dominance of computers, wedded to iron-willed bureaucracy, has completed air transport's faceless personality. Commercial aviation has grown up and, in one sense, died.

Before this time of full maturity the remarkable differences between aircraft caused some to be cursed and others to be loved even though they were contemporaries. There were only a very few aircraft without several faults and even fewer which all pilots agreed were completely trustworthy. Whatever the country of origin, aircraft types were as much the subject of gossip among airmen as other devotions, with reputations being made or damaged according to the status of the speaker. Of the few aircraft with spotless renown was the DC-3. Another was the DC-4, a direct descendent.

The executive dining room at Douglas Aircraft's now forsaken Santa Monica spread was a distinct contrast to the simple office once so continuously occupied by the laird of all the Douglases—Donald Senior himself, a genial, direct man who usually appeared somewhat perplexed at the pomp and circumstance of the dining room. The decor was intended to be club-like, with masculinity rampant, a solid no frills bar, wood paneling, heavy leather furniture, and an enormous, indirectly lit, globe to emphasize Douglas's world-wide affairs. Here the international mighty were invited to dine, ostensibly as a gesture of hospitality and incidentally to whet their interest in Douglas wares.

After refreshing themselves at the bar, guests were escorted to the great round table which would well have suited King Arthur and twenty of his men. Throughout the feast the laird and his knights presided as warm hosts while extolling the virtues of their historic line. Just in case their message was missed, or a language barrier existed, prideful reference was made to the family shields which adorned the room's extensive walls. The display consisted of a superb collection of model aircraft, all rendered in minute detail, all expertly lighted, and all, naturally, bearing the name Douglas. Among them was the DC-4.

An examination of the DC-4's genealogy reveals a lineage without a hint of bastardization. The ancestral mother was the Douglas DWC, child of a 1923 marriage between the original Douglas *Cloudster* and a buxomy float plane known only as the DT. The DWC was eventually betrothed to the C-1, which was directly related to the M-1, 2, 3, and 4, all of pure Douglas blood. The DC-1, DC-2, DC-3, DST, and C-32 were the progeny of that union. These in time became a military family of noble repute, the bombers B-18s and the B-23s.

Now in 1936, with heritage still unmixed, a very large aircraft appeared on the family tree—still pure Douglas but a considerable departure from anything else bearing the crest of the clan. This was the DC-4E, a plane often and erroneously confused with its own descendent, the stalwart DC-4.

Only one "E" was built. It was sold to the Japanese who, after Jake Moxness, the Douglas check pilot, had departed their shores, either forgot or ignored his warning. They attempted a full flap landing with one engine inoperative and promptly crashed, killing all aboard. The DC-4E had a 145-foot wing span in comparison with the DC-4's 117-foot wing span; and a triple tail for easy access to low hangars. The differences between the two types were innumerable, and yet in spite of the "E's" predeath cohabitation with the B-23 bomber there remained a great similarity of profile. Perhaps this was because they had a common midwife in the famous Benny Howard, or simply reflected the clan loyalty. Whatever the reason that aircraft known to the Air Force as the C-54, to the Navy as R4-D, and to the world as the DC-4 Skymaster became one of the most admired of all airplanes.

The DC-4 multiplied until it became a large family of approximately 1,200, all of such sturdy and reliable temperament many pilots developed a special DC-4 yawn. As an instrument aircraft they were so docile even low-time pilots could shoot a near-perfect approach. If one engine failed even at takeoff the effect was nicely manageable and if two succumbed during cruise the aircraft remained very airkindly while producing a minimum of perspiration on crew brows. With nose wheel steering governed by a large hand wheel just above the captain's knee, taxiing was easy and cross-wind takeoffs firmly begun. Ordinary takeoffs were nearly automatic with the DC-4's seemingly anxious tendency to become airborne.

In time their strength became legendary. One C-54 (DC-4) on a night flight between Newfoundland and the Azores inadvertently stabbed a waterspout. The effect was immediate and appalling—and the reactions of the captain were eloquent.

> I was sitting there minding my own business and won-
> dering where I would like to fly when the war was over. We
> had been in and out of some stuff . . . the usual little front
> that lies in there about half way to Gander. It never amounts
> to much. We had been in the air about three hours and the
> crew were moving around the flight deck like they always do
> when they first start to get restless. I had my seat belt tight-
> ened, but not pulled tight. Suddenly it began to rain like I'd
> never seen it before. I tried dimming the cockpit lights and

looking out, but couldn't see a thing. Then it happened . . . I'll never be sure just what or how. I never had a chance to touch anything. . . .

It wasn't very rough air, but all of a sudden I found myself hanging by my seat belt with the buckle cutting into my stomach. It was wild. I was looking up at the floor, papers were sailing past me, and there was dirt in my mouth and eyes. I saw the other guys sort of scrambling around on the roof and I thought sure as hell someone is going to hit all four feathering buttons and that will be it. I hadn't been on my back in an airplane for years and all I could think of was how easy we got wrong side up. And without any real warning except the rain roar which was awful. . . .

. . . while I was struggling to get *up* to my seat I distinctly saw the airspeed needle pegged at 300 and I thought, well, this has got to be it and I'm sorry.

I closed my eyes for a second or so and I'm not sure just what happened next. Maybe I did it, maybe I just helped the forces outside. Whatever it was, gravity returned and after a lot of wild confusion we finally realized we were flying at least reasonably straight and level again. I didn't want to risk going on to Gander so I slowed down, hoping to keep the pieces together and turned around for the Azores. About dawn we landed without any further problems. . . .

The problems came later when mechanics began examining a DC-4 that had exceeded 300 miles per hour in an inverted dive, recovered, and somehow remained a unit. There was very little disagreement when harsh daylight revealed the entire fuselage had been twisted out of line with the wings. She had to be junked. Exactly how or why the DC-4 had managed to stay together under such severe strain defied technical explanation. It was easier to pass the performance off as a miracle.*

* Also in the miracle class. Several years later another DC-4 flying over Texas with a full load of passengers also became inverted. Speed greatly in excess of the allowable maximum was again exceeded. To further complicate matters the hapless crew, while tumbling about the roof, inadvertently feathered three of the four engines. Only the copilot, whose seat belt was properly secured, kept his place and his wits. He managed to roll the DC-4 right side up and save the day as well as a great many lives. For several terrible minutes his only ally was the inherent strength of a DC-4.

Soon after the waterspout episode, another DC-4 crew over Newfoundland, on instruments, discovered a fire in the cargo hold. Returning immediately to Stevensville with the smoke making rapid headway in spite of the crew's efforts, the DC-4 suddenly emerged into the clear. Directly below was a hole in the overcast barely wider than the wingspan, but it was over the airport and descent through it would save the precious time required for an instrument approach. The captain decided he had little choice. He put the ship in a near-vertical bank to stay within the confines of the hole, cut all four engines, and shoved the nose down past 45 degrees. And landed.

The crash crews soon had the fire under control and there were "no further problems." Like the other crew caught in the waterspout, these became devout DC-4 men for they knew of no other four-engined transport aircraft which could have made such a descent.

"Overload" stories about the DC-4 are legion.

Through an oversight in cargo handling during the Korean War, one DC-4 flew from California to Tokyo before it was determined that her rather sluggish behavior was due to a burden of double the authorized weight.

Two months after Pearl Harbor ex-racing pilot Benny Howard, a wiry, no-nonsense man, gifted with near genius in aeronautical engineering, made the first DC-4 test flight. Over the grumblings of his vast engineering staff Don Douglas had allowed the obstinate Howard to make the DC-4 his personal baby, a situation which might lead to disaster in ordinary aircraft production, but in this case it was no ordinary aircraft and Howard was certainly not an ordinary man. The wonderful DC-4 and its separately built military version was not created by a committee. For the first time since the development of the DC-3 in which Benny Howard had also played a considerable hand, both commercial and military customers came in throngs to sup and dine at Sir Donald's round table. Then with remarkable speed, considering a full-scale war was in progress, the face of international air transport began to change. Makeshift bomber conversions such as the Consolidated C-87s, LB-30s, and other sorry hybrids were phased out as rapidly as they could be replaced with a real airplane. Lockheed's

Constellation, the only apparent rival for the DC-4, was still having birth pangs. Due partly to Lockheed's preoccupation with combat aircraft its actual production would be long delayed.

The immediate wartime benefits brought by the DC-4 were spectacular, and for a refreshing change the brass sent the right item to the right job. Even before hostilities became vicious again after the "phony" war in Europe, all realists knew that any invasion of the continent would mean the deaths and terrible hurtings of countless young men. Many Americans thanked whatever God they favored that the DC-4s were ready for D-day. For the dead, when their remains could be found at all, there was only brief and impersonal ceremony, but for the torn and twisted and still living there was sudden and total disillusionment—the forlorn realization of vulnerability. "I'm hit! It has happened to *me*."

Loss of morale for the wounded was as serious as loss of blood, and in some cases more so. Early evacuation from combat dressing stations was a magical cure performed mostly in Douglas C-47s. But the Americans, far from home, were little encouraged by the prospect of a long recuperation in England. So began one of the most inspiring dramas of World War II, the great and swift return of American wounded to their homeland.

Except in the rhetoric of politicians few men willingly offer a piece of their body to their country. The limbs, noses, teeth, and innards are taken from them, usually amid wild confusion, and the owner of the surviving parts sorely needs a restoration of personal faith. The armada of air-evac airplanes eventually served as a primary resurrection in all theaters of the war, and their contribution to total morale can never be calculated.

After the invasion of North Africa, Italy, and Normandy and the various Pacific engagements, the loathsome finale of martial glory could be found in any litter ship. The C-54A carried 24 litters and the later C-54B squeezed in 28. No matter what battle they had survived the passengers were unquestionably the most enthusiastic ever transported by air. They came bleeding and groaning to the wide door, their litters maneuvered with surprising skill and gentleness by medics who obviously gave a damn. Once in position, two and three above the other, bottles to catheters were hung where needed, intravenous feeding tubes snaked from supply bottles, and here and there a bottle of blood was hung so that its contents seeped into a brown mound in the brown blanket below. The cabin became a separate world of bottles and brown people moving quietly among multiple tiers, each supporting a brown mound.

When the door was closed at last there were two brown clad nurses and one brown uniformed sergeant medic who minded the litters. Sometimes they were helped by "ambulatories" who were torn between gratitude for the minor damage to their carcasses and dismay at suffering anything at all. Even after the engines were started no one said very much and the brown mounds were so still the effect was that of a depot in which various covered items were stored.

The agonized eyes all clearly sent the same messages. "Are we really going home?"

Until the C-54 became airborne the young eyes moved cautiously, first up at the sagging litter above, then all around, suspiciously, uncertainly, but still holding a flick of hope in suspense. "How long will it be, oh God? How long?"

Already it had been an eternity.

Litter ships had a special stink of which no one in the cabin or on the flight deck thought to complain. It was the mixture of wet leather, soggy uniforms, mud, and of bodies long unwashed. There was added the bite of antiseptic, the sick-sweet of unguents, and the staleness of plaster bound in the goo of adhesive. To all of this were joined the fumes of excrement both wet and dried, plus the astringent musk of urine accumulated and released. And finally there were some who detected an overwhelming air of indignity, an overall putrefaction of the human spirit which has always attended fallen glory.

There was far too much composure in the cabin until takeoff. It was the indecent silence of private despair.

Transatlantic litter ships normally departed Prestwick in Scotland, and later Orly at Paris, during the early evening. Soon after takeoff, even before cruising altitude was reached, a transformation began in the cabin. The brown mounds stirred experimentally, testing the truth of their predicament. Then almost furtively, as if they feared betrayal, a few dared to communicate.

RENDEZVOUS AT SEA
Douglas DC-4 buzzing Lurline in the mid-Pacific

"Are we over the ocean?"

"Yes. You're going home. Really."

"Goddamn . . ."

Hours passed while the C-54 drummed along between the stars and the sea. The stench in the cabin compounded until it seemed a solid enveloping the brown world which created it. But no one cared. Only the near dead, lost in opiates, slept. Little by little the others came to understand their fortune and with belief came a powerfull revival of themselves. Their eyes became alert. Their voices, ever more frequently heard, became vibrant and sometimes one who had managed to cast off all doubt ventured an incredible laugh.

"Going home? Goddamn!"

By morning of the following day when the brown mounds had been assured they were over United States territory and the fact had been repeatedly confirmed by the ambulatories who kept their noses pressed against the windows, the metamorphosis was complete. All but the most severely hurt were laughing and scratching, buoyed into excited anticipation by the mere proximity of home. It seemed inconceivable that less than twenty-four hours before, they had been in heated combat with an enemy that had already become less than real. Here was a holiday of holidays, a miraculous day never to be forgotten.

Only three years later the same type aircraft is employed in quite another sort of holiday. Now designated as a DC-4 regardless of its military history, several are owned by the Matson Steamship Company, a diverse corporation long engaged in Pacific Ocean commerce. Where once the wounded lay in dubious suspense there are now rows of comfortable seats and the decor of the cabin is quietly elegant. Above the windows there are full-length panels of mahogany veneer expertly carved with the familiar Matson insignia. The stewardesses, obviously selected for eye appeal and wearing uniforms created by the renowned Parisian designer Mainbocher, help the thirty-three passengers become acquainted with generous pourings of vintage champagne, compliments of the line. The passengers' shoulders are still speckled with the confetti thrown at the *Sky Matsonia's* "sailing" from the airport of Oakland, and here and there serpentine still trails from their seats. For Matson is a steamship line with traditions to uphold plus an anxiety to provide the finest for passengers. The officials are still perplexed by the airline business and see no reason why they should not offer the same luxury found on their ships. To this intent they have provided a long table in the rear of the cabin. Seating eighteen at a time the first sitting for dinner will be presided over by the captain of the *Sky Matsonia* and the next by the first officer. Those not actually dining may wander the DC-4 as they please, including the flight deck where they are always welcomed.

The cuisine is superb, created by M. DeGorak, formerly chef to the Belgian king. As aboard the Matson surface ships, the meal will be served by a steward with napkin properly folded over forearm and the menu this night offers an assortment of relishes, bluepoints, and Dungeness crab, with a pleasant Riesling to stimulate the taste buds. Next there will come a brace of guinea hen for each passenger complimented by a generous helping of wild rice and a mellow Cabernet. According to the diner's wish the entrée may be accompanied by a chef's salad followed later by a variety of cheeses, or a generous portion of the *pièce de résistance*—a baked Alaska cleverly molded into the shape of the very DC-4 in which the passengers are now Honolulu bound.

This flying in the grand style is deliberate and the calculated results are most satisfying to Matson officials. They are a rich company feeling their way into a new form of transportation which reports indicate might one day make their luxury liners obsolete. The DC-4s are ideal aircraft for the job, and top-notch crews have been lured from the airlines and the military to do the flying. In all of this endeavor Matson is not alone. A good many capable and well financed transportation companies are trying wings for the first time and are determined to be more resolute than the ill-fated American Export Steamship Line which abandoned its transatlantic air efforts just when it seemed to have a toehold. Even the Grace Line, feudal lord of sea transport in the western Caribbean and coast of South America, has bought a few DC-4s which they propose to operate independently from their quasi-official partner, Panagra. The Southern Pacific Railway has enlisted several DC-4s in an aerial freight service. In England, Cunard has taken to the air.

Unlike the steamship companies the established airlines in the United States are of divided opinions about the wisdom of over-ocean travel. American Airlines, although one of the first to fly the Atlantic, absorbed American Export Lines and then haughtily turned its back on all foreign flights except to Mexico. There was no decent revenue in overseas flights in the foreseeable future. United, among the first to fly landplanes across the Pacific to Australia, was similarly indifferent to foreign operations and was still unenthusiastic about their authorized route to Honolulu.

In contrast, TWA with ten DC-4s in international service is dedicated to offering a complete around-the-world service and be damned to the Pan American torpedos. Their expansive philosophy persists in spite of an executive musical chair game, extraordinary even for the airline business. After a plethora of presidents and vice-presidents were hired, fired, and retired, Ralph Damon, an airman who really knew what he was doing, took over and with eighteen Constellations augmenting the DC-4 fleet he managed to take the limp out of TWA.

All of the neophytes to the flying business were enthusiastic and encouraged by their primary efforts. The Matson Line was particularly pleased by the public response to their monopolistic service. Passengers could fly one way and make the return via a Matson steamship if they pleased. And once arrived in Honolulu they could stay at a Matson hotel. A single ticket covered the entire tour. Everything was luxury class, and yet the price was reasonable. Added features always included cascades of leis, a welcoming serenade from the Matson band when disembarking at Honolulu airport, and often a fascinating rendezvous with a Matson ship somewhere between California and the Hawaiian islands. Forewarned by radio, passengers aboard the *Monterey*, the *Mariposa*, or the *Matsonia*, would line the decks, vying to be the first to spy the approaching *Sky Matsonia*. Once in sight the fun began with wild cheers, handkerchief waving, whistle blowing, and a roaring buzz job by the aircraft. The passengers, both aloft and below, loved it—alas, too much. For every seat on every flight was sold weeks in advance.

A man named Ralph O'Neill, who as far back as the late twenties had organized the very ambitious New York, Rio, and Buenos Aires Airline, could have predicted the outcome. From bitter experience he was aware that good airplanes and good service were only part of the battle. He knew only too well how Pan American regarded trespassers in their domain and how Juan Trippe shot from the hip.

O'Neill need only have watched the Honolulu arrival of a Matson DC-4 and compared it with a similar arrival of a Pan American flight out of San Francisco. The festive crowd disembarking from the Matson aircraft was obviously all she could carry. The PAA flight was often half-filled. Watching, O'Neill must have predicted what all the candidates to the over-ocean flying game still failed to recognize. He would have known that if Pan American could not swallow them as it had his lamented NYRBA airline, other means would be found to eliminate them as competition.

Less than two years after Matson commenced their aerial operations, Juan Trippe arranged a hanging for those he considered formidable outlaws —namely Grace, who must swing in spite of their mutual ownership of Panagra (the steamship line was becoming too big for its aerial breeches), and Southern Pacific Railways, who must also abandon air operations since they were similarly categorized with the most dangerous villain of all, the Matson Lines. Trippe's emissaries pointed out to the Civil Aeronautics Board how each and all of these upstarts were violating the law—specifically the Civil Aeronautics Act which specified that no surface carrier could be engaged in scheduled air transportation. Their guilt was undeniable, and within weeks even Matson began to withdraw from the scene. The Pan American complaint maintained with considerable justification that with all its wealth and outposts throughout the Pacific, Matson might have given PAA intolerable competition in the Western Hemisphere. How much this geo-politicking benefited the taxpayer and the traveling public will never be known, since all the surface carriers were soon out of the air business.

Still the DC-4s were far from shot down. Several factors combined to maintain their status as a valuable and extremely versatile aircraft. Even as they were being phased out by the large domestic U.S. airlines in favor of the more advanced DC-6, the DC-4s found an immediate home with the fast-growing tribe of *non*scheduled air carriers, who through clever footwork managed to keep flying in spite of all opposition. Among these were Orvis Nelson's Trans-Ocean, American Overseas Airlines, Seaboard and Western, Trans-Caribbean, California-Eastern, and the Flying Tigers. Use of DC-4s by those companies based on the West Coast was greatly spurred by the Korean War. One TALOA aircraft went full circle having started life during World War II as a litter ship, then served for two years in Matson's glamour fleet, and finally returned to its original role in bringing home the maimed from Korea. Later the Berlin airlift also depended mightily on the weight-carrying ability of innumerable C-54s and DC-4s. During that crisis, which included some of the hairiest instrument flying ever, one of the most valuable cargos transported by these once-glistening aircraft was coal.

Many DC-4s are still flying the humbler aerial tasks in various parts of the world, and a unique device guarantees their memory will long survive even after the last of the breed goes to scrap. Tucked away in closets and attics of former DC-4 crewmen are countless rolls of paper money. Some are as long as fifteen feet and are composed of the standard currency bills of many lands taped end to end. Known as "short-snorters" for reasons which still remain obscure, these rolls contain as many signatures as their far-roving owner could obtain. They were a caprice of World War II, most popular just when the C-54s were in their prime, and as such represent a montage of people involved in their operation. While most "short-snorter" rolls bear only the signatures of various Kilroys who dropped back into obscurity once they returned to civilian life, the more dedicated collected a composite list of scribblings which in their special way tell of great events and the personages involved. Perhaps with an eye to history they asked for and were obliged with the marks of Eisenhower, on the way to Casablanca; Bradley, Patton, and Clark returning to state-side triumphs; Franklin Delano Roosevelt en route to Yalta; Douglas MacArthur bound for Tokyo; Secretary of State Hull to Moscow; Churchill to Ottawa; General George Marshall around the world; Marlene Dietrich to North Africa; Martha Raye and Bob Hope bound almost anywhere American fighting men were stationed.

The large majority of those signatures were rendered in or about one of the great aircraft of all time.

Chapter Twenty

Lady with a Past

*V*isitors to Israel's Tel Aviv airport in the early 1970s were astounded to recognize the unmistakable conformation of certain Boeing aircraft squatting in the boiling sun. Ten years previously at Oakland airport in California it had been much easier to believe a rank of the same type of aircraft marooned in the grass off the main runways. They presented a sorry spectacle, all sheen gone from their hulls, an entire engine missing here and another there, various elements of the empennage disappeared, windows purpled in the sun, and a general air of shabbiness hanging like netting over the assembly. As the afternoon winds off San Francisco Bay whispered through the crevices and gaps where essential elements of these aircraft had once been and month after month the rains and salt air perpetuated their subtle attack, only the airport jack rabbits brought movement to those neglected monuments. For the passing aviation world was much too busy with jets and a multitude of new electronic devices to care about those fallen queens, who in their heyday had never enjoyed quite the total trust of airmen. It was as if many predictions had come true and harlots were never destined to become grand ladies no matter in what guise they first appeared at the ball. There had been many who believed this final repose in a scrapdealer's alleyway should have been mandated soon after their debut, for in spite of being christened by such illustrious damsels as Margaret Truman and Eva Peron the Boeing 377 Stratocruisers had difficulty living down their earlier sleazy reputation.

A part of this inability to climb the social ladder was timing. In aircraft production, as in anything else, the success or failure of any type is as dependent on timing as on either its attributes or faults. Unfortunately there is such a mandatory time lag of years between original conception and first test flight that not even the most confident swami dares predict the sundry world events and future situations which might influence those customers who should be reaching for their checkbooks.

The Stratocruisers at Tel Aviv airport were a part of the Israeli Air Force and bought at a considerable bargain in relation to their original cost. The Stratocruisers at Oakland were bought at even more of a bargain, specifically $105,000 for a total of 14 aircraft. At the time of purchase four were serviceable, two of which had current airworthiness certificates. The balance were partly cannibalized, and as late as 1972 the bones of one were still in the same place and unburied. As such it represented the last visible remnants of the intrepid Orvis Nelson's Yankee trader enterprise, TALOA, a swashbuckling outfit which included the anything-for-a-dollar Trans-Ocean Airlines. The almost immediate bankruptcy of TALOA after purchasing the Stratocruisers could have been interpreted as further proof that there might have been something of a curse pattern hovering about the careers of these once imposing aircraft.

At Boeing's Seattle spawning ground 56 Stratocruisers were eventually built with Northwest Airlines a customer for 10. They were also purchased by United Airlines who proceeded to phase them out as expeditiously as possible, American Overseas Airlines, and BOAC who operated their all tourist "Coronet" service and luxury "Monarch" service with the type.

Pan American Airways was also an early champion of Stratocruisers, although with a loss of four through one cause or another they were soon acutely aware of whatever jinx kept lurking outside the hangar doors.

Boeing has ever been recognized as creators of extraordinarily stout, admirable, and expensive aircraft. The company has often carried quality to the extreme and customer pleas for corner-cutting, and therefore price tag reduction have been ignored with unusual fortitude. It is sometimes possible to tell a Boeing-built part from an identical part built elsewhere simply by the quality of workmanship and if there is ever any doubt a glance at the price tag should settle all speculation. Unfortunately, fine and honest workmanship does not always eliminate the unforeseen.

On the ground the Stratocruiser was a buxom and quite clumsy looking bird which gave the impression of being much heavier than it actually was. Yet in the air it became curvaceous and graceful, generally acknowledged as one of the most beautiful transports of all time. An early ancestor of the Stratocruiser was the Boeing 307 "Stratoliner" design, a monstrous pressurized tail-dragger often referred to as the "Fat Cat." It too was unusually plump for its time and might have become one of aviation's milestones if again the timing had been right, which it was not. The Stratoliner carried only 33 passengers despite its four engines and cruised no more than 230 miles per hour. TWA had barely time to promote them in their service when World War II turned all energies away from the production of commercial aircraft. The Boeing company devoted itself to producing the enormously successful Flying Fortress (B-17) and later the considerably less faithful "Super-Fort" (B-29).

The Stratocruiser was basically a civilian version of the B-29 with similar wings and empennage. Both types were initially burdened with engine problems, the B-29s with the infamous pyromaniac 3350s and the Stratocruisers with the much more reliable but still temperamental Pratt and Whitney 4360 "corncob" engines. The long list of woes attributed to those 28-cylinder monstrosities defeated the most determined efforts of engineers and public relations officers. Pilots, at first enamored of the roomy 15-feet-by-11-feet flight deck and beholden of certainly the finest visibility of any transport aircraft ever built, soon learned there was a hidden penalty for such an exalted environment. For a few their enthusiasm turned to bitterness and their sense of being master of all they surveyed from such a comfortable throne aloft curdled in response to a procession of unnerving events. The evil engine-propeller combination definitely soured the view from their beautiful domain. Passengers, however, were most favorably impressed with Stratocruiser comfort. The figure 8 shaped fuselage embraced the latest appointments for travel comfort, including berths on long flights.* The lower part of the eight was ideal for baggage and cargo storage and also contained a 14-seat lounge-bar which was nearly always filled to capacity.

Most Americans are relatively carefree as the decade of the 1950s begins. Truman is still in the White House, God is in his heaven, and the U.S. share of world trade stands at the highest it has ever been. Moral values are about the same as the shock of World War II left them, but now a new involvement in Korea labeled a "police action" promises further erosion of accepted standards. A style of music known as "Rock" is just beginning to be noticed, and something called the military-industrial complex has commenced a voracious march through the national treasury. In 1957 when this turbulent decade has just begun to fade, the Russians will launch Sputnik I and thus inaugurate history's longest and most expensive air race.

Moving into the spotlight aeronautical stage-center at the beginning of this decade is the Stratocruiser. It promises to give passengers a whole new way of life aloft and will succeed admirably in this intent—*most* of the time. There are, however, teething problems involved in employing any new type aircraft, and in the case of the Stratocruiser several molars prove to be rotten.

Before anything else can go very wrong, even before airline comptrollers can juggle their magical seat-mile figures, it becomes painfully obvious that the Stratocruiser is an extremely expensive aircraft to operate. The principal American users, Pan American and Northwest, soon deplore their bookkeeping and once again plead for comfort from their rich Uncle Sam. As a consequence a sort of aeroanutical Watergate is revealed. The usual cast of politicians including one Harry Truman, then president of the United States, combined with certain government agencies and a mighty corporation to make the first Stratocruiser operational and keep the rivet guns hammering on more.

Since the price tag on these revolutionary flying machines was heady

* Made up, but rarely sold. The other airlines via their own tightly knit ATA objected strenuously to such "unfair competition."

even by Boeing standards it should have been obvious to the most naive executive that a medium-sized airline such as Northwest would have trouble luxuriating in such equipment unless some sort of financial transfusion could be arranged. The dilemma was conveniently solved by applying an age-old remedy—power plus money. Northwest was invited to apply for a route to the Orient and also Hawaii which would command a healthy subsidy. Next, an arm of the U.S. Treasury, the RFC (Reconstruction Finance Corporation), was induced to provide a suitable loan to Northwest even though the less discreet RFC officials protested mightily and declared the entire transaction "financial folly."

President Truman was not an ungrateful man, and he was pleased that the Boeing company had demonstrated both wisdom and foresight in contributing generously to the Democratic party. It was not surprising then that the politically oriented and appointed CAB found reason to smile upon Northwest's application, make the very necessary award, and thus incidentally assure the sale of ten Stratocruisers.

The key reason announced for such largesse was quoted from CAB policy: " . . . the underwriting must be predicated upon the attainment of optimum costs and revenues and must serve the developmental purposes of the act"—or in less fancy language the CAB was frankly sponsoring an aeronautical experiment which might have been regarded as fair and beneficial if other factors had not been involved. For also included in the CAB dictum was: "underwriting fully the extra cost for each airline that bought the Stratocruiser depends on the wisdom and management in buying the plane and continuing to operate it—not on whether the aircraft turned out to be a disappointment."

Since such logic suggested that the Mad Hatter must be in charge of CAB decisions it followed that Pan American, who had also purchased a bevy of Stratocruisers, should likewise move to the trough. Pan American appealed for a higher mail and subsidy rate on its North Atlantic route to cover the extra cost of operating their magnificent new land version Clippers. Immediate howls of inequity were heard from TWA who competed over the same route with Constellations. They cried with considerably more logic than any other party had displayed that TWA was being penalized because they had been intelligent enough *not* to buy Stratocruisers in the

TOWARD A SMALLER WORLD
Boeing 377 Stratocruiser cruises through a golden afternoon

first place. At this point, while the odor of political clout continued to blight the mobility of Boeing's new transport, other more tangible events contributed to the Stratocruisers' questionable future.

While nearly all pilots were soothed by the ego-pleasing dimensions of the Stratocruisers they were less than enthusiastic about certain of its characteristics once in flight. Once the CAA had been convinced the wing spoilers they had insisted be installed did more harm than good and the devices had been removed, "nose-wheel" landings were no longer necessary. Thereafter, when all went well, the Stratocruisers were easy to fly, land, and take off, the overall skill demands being far less than possessed by those pilots actually engaged. Airmen were beginning to regard them as a forgiving airplane. Even so it was a very good thing no amateurs were at the helm for frequently all things did not go well.

During the initial states of Stratocruiser operation the "corncob" engines displayed a discouraging need for changing cylinders almost as frequently as spark plugs, and the propellers were four villains dancing in a row. Those used by Northwest Airlines had a steel shell with a mastic core filler. If the filler came loose blade breakage came next with spectacular results. The "corncob" engine mounts were of magnesium which instantly gave way followed quite as instantly by the departure of the entire 28-cylinder engine from the aircraft. If, as has since been claimed, this design was deliberate, then the theory was disgracefully ignorant of aerodynamics. For once one of those huge engines had parted company from the mother aircraft the fat was literally in the firewall because its bare face offered such an aerodynamic blockade the Stratocruiser simply could not maintain altitude. Only one runaway propeller mandated an immediate landing.

Northwest was plagued with a series of such failures, yet fortune was so in their league that all occurred within descending range of an airport.

Pan American was less blessed. The Clipper *Good Hope* was lost over the Brazilian jungle, Clipper *Romance of the Skies* vanished in mid-Pacific, Clipper *United States* ditched off the coast of Oregon, and Clipper *Golden Gate* took a blade of the number three propeller through the cabin during a landing approach at Manila. In one of the most beautifully executed ditchings in aviation history Captain Richard Ogg eased his Clipper *Sovereign of the Skies* into the ocean near a Pacific weather ship and all twenty-four passengers plus seven crew were rescued.

There were other slightly less hairy deficiencies in this beautiful bird. Rudder control was inefficient in comparison with those on other aircraft of comparative size. Most of all, Stratocruiser pilots were extremely wary of cowl flap settings which were controlled by the engineer who often became the busiest man on the flight deck. Cooling the "corncob" engines was ever a challenge only partly solved by the employment of over-large cowl flaps. When even partially opened beyond the "trail" position their mass was enough to cause severe buffeting and a very serious loss of lift. As a direct result of cowl flap setting one Stratocruiser was ditched in Puget Sound soon after takeoff from Seattle. When fully loaded, performance on any three engines was always marginal and the necessary wider cowl flap settings to cool the surviving engines usually combined to make the situation even more critical. As a result of such power difficulties, the Stratocruisers carried "Chinese television" on the flight deck, an engine-analyzing device which enabled the flight engineer to observe ignition performance via a cathode tube. The resulting information was most useful in resolving mixture settings, detection of possible valve failures, and in effect reported on the compression in each cylinder.

The flight deck of the Stratocruiser was serenely quiet even if on a good day, just outside the windows, all 112 cylinders were in a reciprocating mood. But that commodious cell had its own peculiar disadvantages. In tropical climes the large areas of glass made it too hot and in colder regions prudent pilots carried raincoats and hats since they could be reasonably certain that soon after descent for landing was begun they would be sitting in the middle of a shower not indicated on their weather charts. The cause was soon discovered, but since the solution demanded the passengers stop breathing the problem was never entirely eliminated. Vaporous moisture expelled by the passengers during their ordinary life process rose and condensed as ice along the stringers at the top of the fuselage. Once the Stratocruiser assumed a descent altitude and passed through the freezing level, the ice melted, flowed forward in rivulets, and eventually emerged as a light rain condition directly over the pilots' heads.

A far more serious and tragic situation occurred on Northwest which was not directly the fault of the Stratocruiser but was a by-product. And here the ravens of political machinations came hungry to the feast. Hoping to avoid a Congressional inquiry into Northwest's Stratocruiser route award and purchase arrangements, the CAB demanded almost impossibly high load factors. Northwest had no choice but to fill them. To perform the job a new and fast twin-engined transport, the Martin 202 was put into service specifically to feed the longer-range Stratocruisers. In later years the Martin 202 proved itself to be an efficient transport, but Northwest, like Faust, having sold its soul to the devil, simply did not have time to carry out a proper "de-bugging" program so necessary to every new type. Nor could they devote the enormous maintenance man-hours required by the Stratocruisers and have money, parts, and personnel remaining to properly service their other aircraft. During the last ten months of 1950 they lost five Martin 202s in a catastrophic series of crashes, which in light of later experience with the same aircraft by other airlines, could hardly be blamed entirely on the Martin Company's product.

The ill-winds of the Korean War and time itself eventually came to Northwest's rescue. The line was heavily engaged in the Korean airlift, and the Stratocruisers were very much liked by paying customers whose only indication there might be something less than perfect about the aircraft were the not so occasional "delays" for "minor maintenance." And in spite of Pan American's uninspiring record, Northwest kept most of their Stratocruisers in operation for more than ten years, many retiring with as much as 30,000 flying hours. Moreover, after so much use, Northwest managed to trade nine Stratocruisers off to Lockheed as down payments on new Lockheed L-188s for $390,000 each—which was very good business indeed.

In spite of its checkered career the Stratocruiser did represent a bold step forward in air transport and fundamentally its subsidizing from the public purse was lawful if unwise. The Stratocruisers showed an impressive survival record with the large majority going to the scrapdealer rather than into jungle or drink. A few in considerably altered form are still alive and well, if the union of many odd parts salvaged from others and combined with considerable new construction can be considered at least a direct descendent of the true blood line. These are the aptly named "Guppys"—Pregnant, Super, and Mini—which hauled Saturn space rocket and Apollo components for NASA. Watching one of these behemoths make an ascension is enough to convince the most skeptical that old Stratocruisers never die.

Chapter Twenty-One

The Little Ones

STINSON "GULL WINGS"

Hard by the Canadian border and off the Washington coast there lies an archipelago known as the "American San Juans." Twice a day ferries maintain the principal link between the islands and the U.S. mainland, but the mails have long since been delivered by swifter means. Every early morning, except for postal holidays, a specter from the golden age of air transport thunders a reveille for island residents as it takes off for the daily mail. Islanders set their clocks by the initial growl of the 450-horsepower Wasp with which the Stinson Gull Wing (SR 10-E) is now powered. In winter the Gull Wing takes off long before dawn and the weather is usually terrible. Winds of 40 knots are not at all uncommon while visibilities vary according to the amount of horizontal rain. A 200-foot ceiling is considered fair flying weather.

About the time the island farmers are finishing their first quota of morning chores, the Gull Wing is returned with the mail. Most islanders take its departures and arrival for granted unless they happen to become involved while it is performing its secondary role as an ambulance plane.

Even Roy Franklin, pilot-boss of San Juan Airlines, has lost track of the number of about-to-be mothers rushed to the mainland. Hundreds of injured and sick people owe their lives to the survival of an airplane which by any standards except its careful maintenance must be considered a flying anachronism.

The Stinson Gull Wing continues as a beautiful example of those smaller transports which in another, more innocent era, excited visionaries who believed there was some future in aviation. Now, thirty-five years after its original construction, modified in many features, it still does a daily job while somehow managing to lift the spirits of the young men who fly it. Older pilots who remember its heyday are sometimes inclined to become misty eyed and make unkind remarks about the gall of a young man flying an aircraft at least ten years older than himself.

The Stinson Gull Wing has always been an extraordinarily pleasant aircraft to fly. Once a pilot learns to make his initial touchdown on the main wheels and allow the tail to settle gently he is virtually assured of a sensually smooth landing.

Although the Gull Wing was rarely used as a passenger transport by scheduled airlines, it was very much a part of the airline scene from the mid-30s to early 40s. In 1939 an enterprising outfit known as All-American Aviation, firmly financed by sport flying and soaring pilot Richard duPont, began the first rural mail pickup service through 55 towns in West Virginia, Delaware, Ohio, and Pennsylvania. The firm used six Gull Wings to dive on a pickup device invented by Dr. L. S. Adams. A hook on a long line effectively snared the waiting mail bag and arriving mail was dropped on the same pass. The system worked so well damage to the most fragile parcel post was almost nil.

The ever so much larger American Airlines also used Stinson Gull Wings for instrument flight crew training and for route checking during the years when the CAA demanded pilots qualified over a route must make periodic landings at every emergency field between terminals.

The tremendous load carrying ability of the Gull Wing plus its inherent stability in rough air made it a natural for bush operations. The Stinson family of aircraft was very large and prominent in the society of smaller flying machines. There were no "problem offspring" in the long line fathered by the great Eddie Stinson and of them all, including the various multiengined airplanes bearing his name, none surpassed the Gull Wing for rugged beauty and ability to earn its keep.

PITCAIRN MAILWING

Today the windshields of many jet aircraft are made by the Pittsburgh Plate Glass Company. Participation by a glass manufacturer in the aviation business is not new, for as far back as 1927 a Harold Pitcairn of the same company became enamored of flight. The keen interest of wealthy young men everywhere in the world has been important to the progress of aviation since its very beginnings. With their faith and dreams untarnished by dull hangar-sweeping chores they fostered many projects which would never have left the ground without them.

Soon after his original interest in flying was sparked, Pitcairn began to think of operating an airline of his own along the East Coast of the United States, but he was dissatisfied with the aircraft available for the job. Proving he was far more than a rich playboy, he hired one Agnew Larsen, a brilliant aeronautical engineer, and one of the several consequences of their union was an airplane designed specifically to carry U.S. mail. This prototype aircraft was interesting if not bold, but by their fifth model, the PA-5, it became very obvious a jewel had been created. Powered with a Wright "Whirlwind" J-5 of 220 horsepower, the PA-5 had very nearly the qualities of a contemporary fighter in maneuverability, an attribute much cherished by mail pilots of the time who had been pushing clumsy DH-4s and other makeshifts through situations for which they were never designed.

The Pitcairn PA-5 carried 500 pounds of mail or cargo and immediately began establishing a reputation for itself on Air Mail Route 19, which was operated by Pitcairn aircraft itself. Now, the still young Pitcairn was very much in business with all the customers he could handle. The price was right—$10,000, later reduced to $9,800. Among buyers were Clifford Ball Airlines flying Pittsburgh-Cleveland, Transcontinental Air Transport, Colonial Air Transport, and Southern Air Transport. With Pitcairn test pilot James Ray at the helm, a PA-5 won the free-for-all race at Spokane, clocking 136 miles per hour. And the sport version, PA7-S, was one of the few aircraft which would hang together through an outside loop.

Although they had a champion, Pitcairn and Larsen were not content. Eastern Air Transport flying the New York–Florida route became a willing customer for the Super Mailwing, PA-8M, which could carry 600 pounds of mail behind a Wright J-6 of 300 horsepower. While the basic structure of fabric covering over chrome steel remained the same, the cockpit was heated and two passengers could be carried if no mail was available.

By 1930, with the United States aching all over from depression chills, both Pitcairn types were becoming legendary. All of them were flying east of the Mississippi where the weather was the foulest, and certainly the element-battle factor contributed to the sense of companionship which became noticeable between man and craft. There was also something less tangible, an open affection for Pitcairns which pilots who flew them still profess. Their enthusiasm makes Pitcairn's sudden about-face all the more unexplainable, for once his Mailwings were well established he resolutely steered his company away from building conventional aircraft while he concentrated on auto-gyros. Some men find success boring, even distasteful, which may account for Pitcairn's abandonment of his lovely mail planes and his preoccupation with a revolutionary and complex flying device which eventually came to naught. Whatever his reasons, Pitcairn's Mailwings won him a place of honor among aviation's great.

BOEING MONOMAIL

It is possible that no other aircraft built in such small numbers exerted such a powerful influence on the pure configuration of later types. The Boeing Monomail, of which only two were built, became something of a symbol. Here was a whole new world of design which confronted the establishment and announced, "You are no longer with us."

The Monomail was a reflection of a very young and spirited industry, a harbinger of that latter-day America when all the world would fly American air products or copy them if buying was too much for native pride.

The Monomail was essentially a test airplane even though it finally earned its living in scheduled service. It was not only the first to feature a semi-monocoque metal fuselage, but also one of the first to exploit "stressed skin" metal as a covering for all the aircraft's elements. More importantly it had a retractable landing gear which did wonders for its airspeed in cruising. Among other factors going for it was the very special Eddie Allen as a test pilot, plus the nearly incredible Erik Nelson as its pilot-salesman. The combination was as successful as it was explosive. Allen was a superb test pilot, and also a swashbuckler of note. Nelson was a rangy import from Sweden who contrived, by some pact with God and the devil, to join the United States Air Corps. Further aided by both authorities he became one of the pilots on the very first Air Corps around-the-world flight. Either in the bar, in the salon, or in the air there was no stopping this magnificent and voluble Swede whose accent was so thick he was frequently arrested during World War II as a German spy.*

To prove the worth of the Monomail, Erik Nelson flew it from Seattle to New York at an average speed of 140 miles per hour, not a record but very fine for the time.

Nelson's East Coast delivery was more a recognition of practicality than a demonstration. The year was 1931, Boeing Air Transport, soon to become United Air Lines, was struggling to stay alive and was hardly in need of experimental aircraft. The West Coast route was rough and tough, pilots were still flying the mail *through* canyons rather than over them, and the Monomail was a relatively blind and clumsy aircraft through such conditions. It would do better east of the Mississippi where agility was not quite so important.

The Model 221 Monomail carried eight passengers, and the 221A carried six, plus 750 pounds of cargo. Yet the numbers are meaningless in relation to the significance of the overall design which disdained flying wires and introduced a cantilever wing monoplane, with a retractable landing gear and anti-drag engine cowling. These features represented true aeronautical progress which has come down to this day in the jet age. It is ironic that the Monomail was actually too far advanced for the state of related contemporary arts. When it first took to the air reliable propellers of variable pitch were still in the experimental stages, and such a heavy airframe needed low-pitch blades for takeoff and high-pitch blades for fully exploiting its aerodynamic hull form at altitude.

---

* An action which unpredictably either amused or infuriated him since he had by this time achieved the rank of General.

Not long after the Monomail's brief career the knowledge gained resulted in the Boeing YB-9, a twin-engined bomber which proved to be the fastest in the world. More importantly for the future of commercial air transport, the revolutionary Boeing 247 became a direct descendant of the Monomail. Then for the first time an all-metal, multiengined aircraft with retractable gear set a pace for the rest of the world.

NORTHROP (ALPHA THROUGH GAMMA)

Through one of those curiously unreasonable series of events which have always seemed to haunt the flying business, William Boeing, whose own company produced the Monomail, simultaneously engaged his considerable money and prestige in fostering the only airplane which could be considered a direct rival. Whether his patronage was simple generosity or merely protective of Boeing interests is still open to debate. The facts are that it was Boeing who arranged for John Northrop, an architectural draftsman and self-made aeronautical engineer, to join with the very knowledgeable Don Berlin in a newly established Northrop Aircraft Company. Its immediate enterprise was the design and production of single-engined, all-metal monoplanes for airline use.

The need was very apparent. For reasons best known to the bureaucrats in the Department of Commerce, both the trimotored Fords and Fokkers were still restricted to daytime flight. Thus someone in something had to provide the airlines with their bread and butter by flying the mails at night. In their wisdom the Department of Commerce thought multiengined night flying would be dangerous, but it was quite all right for single-engined aircraft. The mail pilot sat on a parachute, knew how to use it, and never mind the effect of descending aluminum on those who might live below.

So the Northrop Alpha was born and fortunately turned out to be a very fine aircraft. It was smaller and much more agile than the Monomail and nearly as fast. It was an excellent small field airplane and relatively simple in construction. Tests proved that when equipped with wheel spats the landing gear created only slightly more drag than a retractable assembly, and the weight saved was considerable.

Although the principle business of both models was mail delivery, the

"ALL IN THE FAMILY"
Left column (top to bottom): Pitcairn Super Mailwing, Northrop Alpha, Consolidated Commodore "16," DH-86; middle column (top to bottom): Boeing Monomail, Blohm und Voss HA-139, Stinson Gull Wing, Farman F-3X (Stork), Lockheed Orion; right column (top to bottom): Piaggio Royal Gull, Savoia Marchetti S-73, Boeing 247, Fokker Super-Universal.

Robert Pacco

Alpha 2 model could seat six passengers plus the pilot while the "3" seated only three in addition to a payload of mail or cargo. National Air Transport (NAT–UAL), apparently content with their Monomails, bought only one Alpha, but Transcontinental and Western Air soon had twelve. Clarence Young, then Assistant Secretary for Aeronautics, bought one for his personal use. The price was initially high, $21,500, but did not hold very well since by 1938 Alphas could be had for a mere $2,000.

Because of its immediate success the Alpha spawned other models, the Delta, which was much more of a passenger aircraft, and the Gamma, in which Jack Frye of TWA set a coast-to-coast record Los Angeles to Newark, 11 hours 30 minutes.

It was in a Gamma during the mid-thirties that the redoubtable D. W. (Tommy) Tomlinson pursued his less spectacular but far more important work in high altitude "over the weather" research. The Gamma (2D) could carry an 1173 pound payload to 20,000 feet and cruise at 215 miles per hour, a remarkable performance for the time. Before his death, Wiley Post had also contributed to high altitude research, but on the whole his investigations were discouraging whereas Tomlinson's more methodical experiments offered great promise for the future. The development and use of pressurized aircraft by TWA and other airlines can be traced directly to Tomlinson's early flights.

Once again corporate maneuverings caused unfortunate ruptures in what should have been a continuously exciting enterprise. In 1931 there was an alarming collapse in the aircraft market as there was in everything else. As an economy measure the Northrop Company was summarily ordered to marry the Stearman Aircraft Company and move into its Wichita home.

John Northrop found himself so unenthusiastic about the match he picked up his slide rule and remained firmly in California where he intended to continue his research with his highly controversial "flying wing" and other monoplanes, both military and civilian.

As subsequent aviation history reveals, John Northrop was a practical radical in his thinking and his measure as a creative designer may be taken from the long line of splendid craft bearing his name.

CONSOLIDATED COMMODORE (MODEL 16)

Until 1930 large flying boats seemed to be the exclusive interest of European countries. There, many designs were created and several actually constructed, including such freak examples as Italy's eight-engined Caproni and Germany's grandiose twelve-engined DO-X. The United States was still considerably behind England and the continental countries in the development of land-based commercial aircraft and lagged even further behind in water-born. Finally the U.S. Navy decided a flying patrol boat might be more handy than lighter-than-aircraft. The answer to their requirement was provided by the Consolidated Aircraft Company which produced the XPY-1, a twin-engined flying boat better remembered as the Admiral. It proved to be a very efficient aircraft with excellent water handling characteristics. Its trials were satisfying to all concerned, but no one could then foresee that less than ten years hence its direct descendant, the PBY, would become one of the most famous aircraft ever built.

The civilian version of the Admiral was the Commodore, a low-slung, sleek flying boat powered by two 575-horsepower Hornet engines. Ralph O'Neill's newly established New York, Rio, and Buenos Aires Airline (NYRBA) had ten Commodores in service by 1930. Flights left Miami once a week and proceeded southeastward in eight easy stages. In spite of the relaxed schedule the Commodores still arrived in Buenos Aires eleven days before the fastest surface ship could make the same voyage.

Commanded by pilots who wore "Sam Brown" belts as well as four sleeve stripes, the Commodore cruised at a leisurely 100 miles per hour, fast enough to impress the twenty-odd passengers and governments who might offer mail contracts, and yet economical enough to assure a range of 1,000 miles.

Mrs. Herbert Hoover, wife of a man who could hardly have chosen a worse time to be president of the United States, christened the first Commodore. Whatever good wishes for its safety she might have offered were apparently effective, for all the Commodores made a splendid record for themselves. Unfortunately she must have neglected to mention Ralph O'Neill, their original sponsor, and wish him long tenure. He was standing in danger, for Pan American, commanded by Juan Trippe, was just testing its own water wings and viewed the NYRBA operation with a mixture of envy and dismay. It was already obvious that if O'Neill and his transtropical flying circus were allowed to continue it could become difficult if not impossible for Pan American to control the Caribbean and South American airways.

Times were depression and becoming increasingly tough for all but the very rich, a factor which contributed greatly to the simplicity of solution. If NYRBA could be cut off at the financial pass, O'Neill would soon fall on his own sword. That pass was a mail contract from the U.S. post office, an establishment which O'Neill would fail to persuade against the winning blandishments of Juan Trippe. Thus in spite of O'Neill's mail contracts with South American countries he was soon outflanked. Pan American bought his ten Commodores at prices ranging between 97,000 and 106,000 dollars.

To ease the pain, O'Neill was offered a job with Pan American which he promptly declined. He was a young man who had a dream which he had managed to make a temporary reality. But he was up against another young man who cherished an identical dream and who happened to be just a little bit tougher.

Once sole proprietor of the southern skies, Pan American bought four more Commodores and continued to fly the type until 1935.

FARMAN F-3X (STORK)

Since aviation's earliest days the Farman Company of France has been one of the world's most prolific and honored aircraft manufacturers. Aeronautical heritage is an important phase in French history for the nation has a long list of pioneers dating back to the days of hot-air balloons and the latter-day exploits of the dauntless Bleriot.

The whole language of aviation still owes much to the French tongue and even the normally monolingual Yankee airman still spices his speech with "aileron," "fuselage," and "empennage."

Except for a very few small experimental types, the Farman Company (Avions H. & M. Farman) concentrated on the building of large aircraft. Surprisingly for a manufacturer that had been in business since before the 1914–1918 war, the Farman people displayed a reluctance to modernize

179

their designs even after combining with the Hanriot Company in 1936 to form the Société Nationale de Constructions Aéronautiques du Centre.

There is a hoary and very sound saying in aviation—"an airplane flies like it looks." If there were not exceptions to every standard, then Farman's F-3X Stork never would have managed its first levitation.

Very possibly the Stork could qualify as the ugliest flying machine ever assembled, although various other Farman creations might have served as competition. In appearance nearly all Farmans project a configuration that suggests the designers took on a full load of cognac before applying themselves to their drawing boards. In the case of the Stork they may have been suffering monumental hangovers when they conceived a high-wing monoplane with four water-cooled engines set in tandem pairs above the landing gear. With little regard for aerodynamics they built a boxy fuselage featuring aquarium-sized windows and a round nose which might well have suited Captain Nemo's submarine. All of the cabin area was occupied by passengers who sat in wicker sun-porch chairs.

In an apparent concession to the idea that a pilot should be included somewhere aboard, the Farman designers provided an open cockpit situated on top of an enormous square-tipped wing. Once seated in his lofty foxhole the pilot's view was obstructed by much of his flying machine. He could see up, but not down, backward for what interest he might have in the empennage, and over the nose only if he forced the Stork into a descending attitude. In contrast, the passengers enjoyed a tremendous view. Passengers who dared open their eyes once this awkward contraption left the ground could at least find some comfort in observing there was one engine for each individual.

Some measure of contemporary French standards may be appreciated by noting the Stork won first prize for safety requirements in the 1923 Grand Prix des Avions Transports.

Not in the least embarrassed by their grotesque product the Farman Company put four Storks into service on their Paris–Brussels–Amsterdam route. Then, after what must have been an historic example of French salesmanship, Danish Airlines bought two and flew them to Copenhagen, Hamburg, and Cologne.

If the F-3X Stork was the Cyrano of French aviation, the F-4X, which sported three round air-cooled engines rather than four in-line liquid-cooled, was the hunchbacked Quasimodo. There were some who maintained that such a flying junk pile could only have been designed by a Frenchman suffering from a very bad liver or unrequited love. Four of these monstrosities were delivered in 1925 and one actually inaugurated the Paris–Zurich service. Mercifully their career was short and the Farman people eventually returned to their senses.

FOKKER SUPER-UNIVERSAL

For certain individuals Berlin's Adlon Hotel was a gay place even during the grim times of 1917 when it was already apparent all was not going as planned with the Kaiser's war effort. Yet adversity often favors the resourceful, a good many of whom could be found drinking the Adlon's excellent champagne at no cost to themselves simply because they were heroes. Temporarily returned from the front they were registered under such names as Richthofen, Udet, Jacobs, Müller, Loerzer, and Goering. Their rooms and the *fräulein* to make them comfortable were all paid for by a Dutchman named Tony Fokker who had made it his business to provide airplanes to the highest bidder—in this case the Imperial German Air Force. The task was not too trying because with the help of Martin Kreutzer airman Fokker designed excellent airplanes which the Jastas were delighted to fly and fight. He was also a generous host who enjoyed a party himself and he knew the value of famous heroes endorsing his product. Allied airmen profoundly regretted that Tony Fokker provided airplanes for what they thought of as the wrong side.

The Allies might have benefited from Fokker's talents if they had understood the young man's ambitions better and had been willing to kidnap him from either Germany or his own country. When many of his Jasta heroes had been killed and Germany collapsed, the Dutchman Tony Fokker did not go down with it. Instead with characteristic deftness of foot he teamed with a designer named Platz and together with another designer, Robert Noorduyn, made his way to the United States. Fokker immediately employed his boundless energies in the production of aircraft for peaceful purposes.

One of Fokker's many designs which achieved flying reality was an immediate success. It was a monoplane he called the Universal which described it very aptly. Equipped with a 200-horsepower Wright J-4 engine, it was U.S. certificated less than seven years after the last 1918 cannon was fired. And Tony Fokker was only beginning.

The first Universals carried four to six passengers plus the pilot. Fokker's unmistakable, no-nonsense style, perfected during World War I, was maintained in the simplicity of the Universal. There was considerable discussion as to whether Fokker had learned this very valuable aeronautical attribute from the Germans or took it to them.* Regardless of credit allocation, Universal buyers saw an aircraft which sold at a reasonable $14,500 and promised not to devour profits in maintenance. Early customers were Colonial Air Lines and Pacific Air Transport, who equipped their Universals with slightly more powerful engines (Wright Whirlwind J-5s, 220 horsepower) and enlarged the cockpit to accommodate two pilots.

The Universals were built at Teterboro airport by Atlantic Aviation, a subsidiary of the Fokker Aircraft Corporation which was already well established in Europe. Although Fokker spent a great deal of time commuting across the Atlantic in steamships he was making money while competitors were struggling or even failing.

After the original Universal a "Standard" was produced which was essentially the same airplane, but featured an enclosed cockpit and a 300-horsepower Wright J-6 engine. It carried eight passengers and still evidenced Dutch frugality by wearing a tail-skid instead of a tail-wheel. A final version known as the Super-Universal made its debut in 1928. It was happily flown by Standard Airlines and Universal Airlines, both of which eventually became a part of American Airways.

By the early thirties Fokker's far-reaching enterprises were in full swing. In the United States alone 40 Universals had been built and 60 Super-Universals. In addition Fokker was building his trimotors using the same basic materials, wood-laminated spruce spars, plywood ribs, and stringers with plywood veneer covering. Fifty trimotors were built and in addition to their airline service set a notable series of flying records.

* A virtue American designers have failed to endow their military aircraft with despite four flying wars of experience.

Ever since youth Tony Fokker's star had remained in the ascendant. In his later years when he enjoyed a purely social game of matching slide rules with Donald Douglas, it was even rumored that his expertise contributed to the DC-1 and DC-2. True or not, Fokker's memorial was more than stone on a plot of ground. His aircraft, particularly the Super-Universal, continued to fly until relatively recent times as one of the bush pilots' most valued friends.

LOCKHEED ORION

During the early thirties a curious aeronautical situation prevailed in the United States. While the large airlines were lumbering along at only a little better than 100 miles per hour with trimotored Fords and Fokkers, their much smaller confederates, New York-and-Western, Continental Airways, Ludington, and Varney were flying the fastest transport in the world.

Once again the Lockheed Orion proved that an airplane flies like it looks. Even the early models powered by a Wasp 450-horsepower engine could carry seven passengers in a pinch and five comfortably at 180 miles per hour. It was a cruising speed not to be matched until the day of the DC-3s.

Although Lockheed has built failures, it has never been guilty of producing an ugly airplane. The Lockheed look has always been a fast and sleek look with the Orion proving no exception. It could achieve a top speed of 210 miles per hour and appeared as if it yearned to prove it even when parked on the ramp. The price was high for a single-engined aircraft ($25,000), but then so was its ceiling at 20,500 feet, an ability which made it particularly useful over the western United States.

The Orion featured what was probably the first sliding hatch over its cockpit and a landing gear which retracted in 30 seconds—faster if the pilot pumped the hydraulic system handle with more than average vigor.

Great names were involved in the development of the Orion, Carl Squiers and test pilot Marshall Headle among others, but no individual has received sole credit for its design. A belated bouquet should be sent to those responsible for conceiving a tail assembly which defied the designers' habitual penchant for drawing it too small. The Orion's tail was right the *first* time, thereby setting a unique aeronautical record.

The Orion attracted so much worldwide attention it was not long before a Swissair delegation arrived to discover what all the applause was about. Impressed with the Orion's performance they paid it the ultimate Swiss compliment—solid Swiss francs for two copies powered by the new 575-horsepower Wright "Cyclone." After the aircraft made a promotional tour of Italy, North Africa, Turkey, and Austria, they were put into regular Swissair service and left every other European airline 60 miles per hour in their wake. In 1936 they were sold to Spain for war duty with the Republican Air Force.

In almost every respect the Orion was a magnificent airplane which exerted considerable influence on aircraft still to be designed. Even so, only fourteen were built, hardly enough to flood the very real market for such performance. It was unfortunate that the Lockheed brothers chose to separate soon after their company was founded. Allan, who died in 1969, remained in a consultant capacity until his end, but Malcolm left the world of aviation resolutely behind in the early twenties and devoted his attention to the invention of four-wheel brakes for cars and other vehicles.* Perhaps if they had remained together longer the Detroit Aircraft Corporation, of which Lockheed was merely a subsidiary, might not have passed its 1931 economic squeeze down to the only profitable part of their operation. Instead of a cut back the beautiful Orions might have lifted the company right through the Depression and with modifications dominated the single-engined transport category.

One other evil matter influenced possible Orion customers. In various record attempts the hollow metal props which pulled the Orion so swiftly across the skies had a tendency to break and start serious engine fires. The inconvenience occurred often enough so that for a time, reaching for a parachute's D ring became known as the "Lockheed Salute."

DH-86 (DIANA CLASS)

While there was a time when the sun never set on the British Empire, the ubiquitous little DH-86 was flying in round-the-clock sunlight long after England's colonials began going their separate ways. A biplane powered by four in-line 200-horsepower engines, the DH-86 was as British as a tea cozy.

* At which he was reputed to have made a fortune.

Sixty-two of the Diana class DH-86s were built and flown by such Imperial Airways stalwarts as Captain O. P. Jones (OBE) and A. L. Wilcockson (OBE). They were flown regularly within the United Kingdom, the special buzzing sound of their four little engines making them seem like swarms of horseflies migrating between London, Glasgow, and the east and west coasts of England. In service with Imperial Airways they carried the flag to the continent and thence to remote lands which had known British rule since pre-Victorian times.

At Kano in Nigeria the arrival of a DH-86 Diana was marked by a seven-foot-tall Nigerian functionary who stood at the end of the runway and blew mightily upon a brass trumpet which matched his own height. With much of Kano's ambulatory population present, drum-beating and dancing ceremonies would commence the moment the aircraft touched down. Then in due time, displaying proper British indifference to the barbaric ceremony and the blazing African sun, the Captain would descend from the aircraft as if he had merely made another landing at Croydon.

The rather fragile appearance of the DH-86 Diana was deceiving. Although the accommodations were spartan she did carry ten to twelve passengers at a cruising speed of 145 miles per hour. Very few 1934 aircraft could match this capability nor could European competitors equal its lightness of control or ability to switch from purely continental flights, Brussels–Cologne–Halle–Leipzig–Prague–Vienna–Budapest where the up and down action became monotonous, to the long hauls over middle Africa or the humid environment of the Penang–Hong Kong–Saigon routes.

In the best pukka sahib tradition Elders Colonial Airways flew DH-86s between Lagos and Accra, British Continental flew them London to Malmo, while Quantas used them in Australia.

Before hostilities in the Far East were declared official, one DH-86 was shot down by the Japanese despite the prominently painted British ensign on both wings. Yet almost as if the dainty little Dianas ignored violence they continued to fly throughout the Second World War.

BLOHM UND VOSS HA 139

The Germans have built some of the world's most hideous looking aircraft and they have also built some of the most handsome. Among their very finest in appearance was the *Blohm und Voss HA 139*, an over-ocean float plane with the flowing lines of a fine sculpture. Visually everything about it was in flawless proportion, even if a certain German stolidity seemed to dominate.

Blohm and Voss were ship builders of Hamburg with a long-time tradition of strength and exactitude in their water craft. When they built the diesel-powered HA 139 for Deutsche Lufthansa, the automatic sponsorship of the German government made it possible for them to proceed in an experimental fashion and ignore economic demands for the sake of passengers. Even so, Dr. Ing Vogt, the man responsible for the HA 139's career, was challenged with aeronautical problems rarely faced by designers. His nation had long-established ties with all of South America and with the rise of Hitler the intent was to make them even stronger through faster communication. Also the race for future North Atlantic air traffic was still wide open, a nebulous market which Lufthansa along with most European airlines saw as lucrative long before U.S. airlines developed anything but passing interest.

In spite of the Zeppelins' success, Lufthansa reasoned that some other means must be found to carry significant payloads over the great ocean distances. The Dornier DO-18, another example of superb flying-boat design by the Germans, had already proved the valuable endurance of diesel engines in long-range aircraft. One flew from the Azores to New York in slightly more than 22 hours and arrived with 10 hours fuel remaining! Yet pretty as they might be, the DO-18s were only two-engined and their configuration was such any future enlargement for passenger work would be nearly impossible. Blohm and Voss's HA 139 was quite another swan. She could be quadrupled in size and still remain the same basic airplane. For the present Dr. Vogt and his associates had to be content with an aircraft that accommodated only two pilots, a flight engineer, and a radio operator, but the potential for expansion was built into the initial design.

While it is inconceivable the Germans thought they could ever persuade even the most rugged passengers to submit to the breath-taking sensation of a catapult launch, the HA 139 was deliberately designed for such strains and actually operated off the depot ships *Schwabenland* and

Friesenland during the initial North Atlantic flights. Through summer and fall of 1937 seven round-trip flights were made between the Azores and New York with an average speed of 150 miles per hour—not at all bad considering the winds along the route are normally light and, unlike those in the higher latitudes of the North Atlantic, usually cancel their effect on two-way flights.

When the trials to the North American continent were completed the HA 139s moved down to the South Atlantic and served on the Bathurst–Natal–Recife service of Lufthansa. The majority of flights were made at normal altitudes, but some German captains preferred to take advantage of ground effect and flew the entire route day or night at less than 50 feet. They claimed to have developed a special technique for remaining hour after hour at such nerve-wracking altitudes. If true, it is little wonder that when they finally sighted a certain uplifted prominence off the Brazilian coast they lit a fresh cigar and saluted it as "the finger of God."

SAVOIA MARCHETTI S.73

No record of air transport's golden age would be complete without some recognition of the Italian contribution. Even before the First World War the Italians were experimenting with aircraft construction and soon after that conflict they directed their very considerable talents to serious production. Powerful names became interested in the brilliant blue skies of Italy as something more than a permanent umbrella over their pasta. Fiat, Breda, Caproni, Savoia Marchetti, and Piaggio were among the several combines engaged in the production of aircraft, most of which were excellent products.

In 1934 the Savoia Marchetti S.73, a trimotored, 18-passenger transport, was unquestionably one of the finest flying in the world. As flown by Ala Littoria they were painted an eye-stunning creamy white with Mussolini's chop, a brightly painted Roman fasces, decorating each side of the nose. In overall configuration the S.73 resembled the German *Junkers 52*, but it was far lovelier to look at and offered more salubrious passenger accommodations. Cheap labor and relatively unpolluted skies enabled the Italians to keep their S.73s immaculate inside and out and if their pilots were inclined to "cowboy" approaches and landings, their performance was endured as merely a reflection of Latin temperament. The bountiful food, wine, and decor forgave whatever tension 60-degree banks may have placed on passenger nerves.

As the *S.73s* gained popularity with European airlines it became apparent this was an aircraft which could be flown with almost any reasonable choice of power plants. Ala Littoria used the Piaggio Stellas as well as Wright Cyclones. Avio Linee Italiane installed Alfa Romeo engines, while the Československé Státní Aeroline equipped theirs with Walter Pegasus IIM2s. Sabena, which flew the *S.73s* from Brussels all the way to Elisabethville in the Congo in 44 hours flight time, used the Gnome-Rhône Mistral Majors.

The *S.73's* wings were plywood-covered. The steel-tubed fuselage was fabric-covered, yet so expertly done it appeared to be metal. In spite of their numbers not many *S.73s* survived World War II.

PIAGGIO ROYAL GULL (L-I)

The so-called Royal Gull was a much smaller postwar aircraft, in many ways typical of Italian design efforts prior to the jet age. Viewed from the front they have often been mistaken for jet-powered aircraft. They were originally intended for the Italian search and rescue service and thus were of extremely rugged military construction. Even so the Italian sense of graceful proportion remains so distinct it is still the only amphibian in the world which can taxi out of the water and onto the beach without offending the eye.

The Piaggio *L-1* and the later *L-2* are a delight to fly off and on the water, but partly because of their short coupling and stiff gear they can make a pilot earn his landing when setting down on solid earth. Their pusher engines afford a very satisfying rate of climb for an amphibian. Rough water takeoffs and landings in a Piaggio are not a problem, the tough hull and wing structure providing an extra margin of comfort to the pilot who cannot wait for smoother seas. And whether by accident or deliberate design the Italians managed to build the first amphibious flying boat which on occasion does not behave like a submarine. If a very short landing is required, the nose of the Piaggio may be shoved full forward immediately after contact with the water. The consequence is an astonishing stop with the nose failing to display any desire to bury itself. If sea conditions are flyable at all it is rare to see more than a few droplets of spray on the windshields.

The little Piaggio amphibians carried only four passengers plus the pilot, but in service they have enjoyed a unique series of very different and quite special environments. In addition to their intended service over the Italian boot, a Milwaukee machinery manufacturer (Trecker) imported 23 Piaggios in the early 1950s and for advertising appeal dubbed them Royal Gulls. Commodore Aviation in California bought several for service between San Francisco Bay and the gambling casinos along the Nevada side of Lake Tahoe. The seating was simplified and rearranged so that six passengers plus pilot could be carried. When passengers were lacking during certain midday flights, full loads of San Francisco newspapers were carried to the remote lake. With summertime temperatures in the nineties and an altitude of 6,225 feet, plus an almost perpetual lack of any helpful wind, full-load takeoffs at Lake Tahoe could become very sticky. The Piaggios accomplished the chore day after day with a minimum of fuss.

Other Piaggio assignments were with Macao Air Transport on their service to Hong Kong. Then a few went to New Zealand and Australia, while several are still flying in the Pacific Northwest. As a peculiarly special token of esteem a certain Aristotle Onassis hoisted a Piaggio amphibian aboard his fabulous yacht, *Christina,* and used it for shore-side transportation.

There remains one irascible fault in the Piaggio Royal Gulls. Pilots who fly them are inclined to wager with the innocent that one wing is shorter than the other. Warning. Do not insist otherwise. It is the Italian way of compensating for power torque.

Chapter Twenty-Two

On the Beak of an Ancient Pelican

My heart had long been scorched with envy, for other men were lofting to regions I could never achieve. It was the year of the Geminis, of plans for the moon, of supersonic transport design, of fighters slashing thrice the speed of sound. Like most people I had no choice but to remain an observer, a grubby role for one who has flown with eagles. Perhaps that is why I instantly agreed when Freddie called and asked, "How would you like to fly a DC-3 from San Francisco to Apia?"

If there is anyone who does not know where Apia is, then it is in Western Samoa, which is very far over the South Pacific horizon.

It had been nineteen years since I had flown a DC-3. Where now was my hard won wisdom? There was the belief I had always held that a wise man never tries to go back?

And yet . . . Apia, a siren whispered the name. An author named Robert Louis Stevenson is buried in Apia and if he could make it in a sailing craft, certainly I should be grateful for a DC-3. The analogy would be abused, I knew, by well-meaning, jet-minded friends.

"A *DC-3*? It's four thousand over-water miles to Samoa!" A preliminary measuring reminded me it was *four thousand, three hundred and fifty miles.*

"You'll go crazy! It will take you thirteen hours just to Honolulu. . . ." *My specially designed pessimistic computer insisted it would take longer.*

"A jet takes only four hours plus. Stay home and write books. No one ever drowned writing and making a fortune."

But how much had they lived?

"What happens if one engine quits?"

According to my recollection most DC-3s eventually arrived at their destination if they carried enough fuel. In my private manual I firmly believed the only time there was too much fuel aboard any aircraft was if it was on fire. As for single engine emergencies, I had enough familiarity with the proper mixture of fright, sweat, and faith to remain convinced "it can't happen to me."

"All DC-3s are ancient. What about metal fatigue? If you take a tin can and bend it a million times . . . well?"

Well? Never having flown with the handicap of an engineering degree, I had never worried about such things. But I would bend as gently as possible.

Freddie, while masquerading as just another Pan American pilot, was, as everyone knew, the uninaugurated president of the Pacific Ocean. On the telephone he had advised, "The father of our country will be your navigator."

I was pleased because *this* George Washington was a stocky, alert, New Zealander, at the moment Operations' Manager-Chief Pilot-all around-high chieftain of Polynesian Airlines. And he smiled easily. This infant airline had been flying a route pioneered by Captain Cook, rechecked by William Bligh, and publicized by Somerset Maugham and James Michener. With a single borrowed DC-3, Polynesian Airlines had been serving Apia in Western Samoa, Pango Pango in American Samoa, flying thence to Atitaki and Rarotonga, or westbound to Tongatapu and Fiji. Now, after two years of operation Polynesian had taken an important step. Business was so good they had resolved to buy an airplane they could call their own. Following sound advice they had bought a DC-3.

Freddie said, "John Best will be Flight Engineer. He can also do some of the flying when you want a stretch."

Best was also a New Zealander. Though still in his early twenties he approached genius as an aircraft mechanic. It was he who had nursed Polynesian's single rented DC-3 so tenderly, soothing its brow against all weariness. Many people believed the line operated four airplanes.

"And who," I asked Freddie, "will be the copilot?"

"Copilots for ferry flights are hard to come by . . . you can sort of switch around."

Freddie is easily given to sweeping statements when bothersome details threaten his multitudinous affairs. He is a man who likes to launch projects. If allowed he will plan your coming week, month, year, or life.

"Freddie," I said patiently, "George Washington is going to be very busy navigating and when he is not actually holding octant in hand he should be catching a few minutes sleep. John Best should be checking fuel consumption and a lot of other things. Without even knowing what the winds will be, it will certainly take us fourteen hours or so just to make Honolulu. That is a long time for these bifocaled eyes to be staring at instruments. There should be a copilot, someone—"

"What about Dodie? She could double as stewardess."

I swallowed thoughtfully. Dodie was my girl Friday secretary. It was true that she was taking flying lessons and was almost ready for her private license, but when she signed on for her job, the fringe benefits did not include a possible voyage in a life raft. Personally, I would feel much safer a thousand miles from the nearest land in a DC-3 than on any freeway, but Dodie's decision to go might hinge on loyalty. I remembered only too well that ferry flights were never the same as routine passenger flights. There would be the usual makeshift arrangement of extra fuel tanks installed for one flight only, and of course a subsequent weight overload. Yet the ferry flights I knew about had arrived at their destinations in good grace . . . almost always.

"I'll ask her."

Thus it was that the fourth member of our crew was a girl named Dodie. Do not offer adventure to a certain kind of female unless you want them to accept.

The San Francisco night is unusually soft, and a near full moon is rising across the Bay. It is Friday the thirteenth, which may have accounted for three lucky takeoffs and landings I had executed during my afternoon reunion with a DC-3. After nineteen years . . . there she stands quite as resolute as ever, a bit paunchy-looking perhaps with the new type landing gear doors, but otherwise the blood sister of those I had flown regularly from New York to Cleveland and Chicago in 1939, to California in 1940, and across the Atlantic to Greenland and Iceland when there were no radios to guide us because the towers for constructing same were our cargos.

Below the cockpit window I notice her christened name—*Savaii*, the name of the second island of the two which constitute the new nation of Western Samoa. John Best has painted the red and blue national flag on her tail.

Beyond *Savaii* is the enormous San Francisco airport. Jets keen their elephantine way along the runways, others sigh down one after another for their landings, still others blast their hot breaths against the night with power we had never dreamed of only a few years ago. I watch them soar toward the moon.

The contrast seems almost too much for the *Savaii*. The brilliant hangar light is cruel to her, the new paint becomes the pitiful striving of an old harridan trying to look her best at a relative's wake. There is something sheepish about her. And why not? In a few moments I will guide this anachronism along taxi-ways five times wider than she requires. The takeoff runway is so long that even with an overload *Savaii* should be able to make an ascension from one end, fly momentarily, and land at the other end with room to spare.

John Best comes to my side. "We are ready."

"All tanks topped off and checked?"

"Personally . . ."

After a few minutes we are taxiing slowly toward a moon path on San Francisco Bay.

George Washington calls the control tower and I try to persuade myself his New Zealand accent is to blame for the patronizing tone in the controller's voice. The tone changes to consolation when he recites our airways clearance to Honolulu. Beneath the obligatory technical mish-mash he seems to be saying, " . . . now, not to worry. But are your life jackets handy?" The coward within me is momentarily resurrected, then dies one of his ten thousand deaths.

When we run up the two engines and check the magnetos we sound like vacationists playing with their outboards on a quiet lake. Just behind us, crouched like a prehistoric monster, is an American Airlines jet. I must know the pilots, or at least the captain. Long ago, beyond the swiftly closing mists of aviation time, we must have flown over the same routes as comrades, in DC-3s which really *were* brand new and glistening, and of which we were extremely proud. And it is very possible the rest of the captain's crew have never flown an airplane with a propeller on it.

I am reasonably sure what they are saying on the flight deck of the American 707 while they contain their impatience with this obstructive gnat, "Some people have it tough . . . flying a beat-up old DC-3 to Honolulu."

Lo, how the mighty have fallen.

Moments later *Savaii* demonstrates that however humble, she is far from beat-up and is not about to join the fallen. In spite of the overload she soars from the runway like a frightened sea gull. As altitude and airspeed mount I yell triumphantly, "We are in orbit!" The night allows me the deception of playing astronaut. And those who have been there have told me there is no more "G" sensation transmitted to their backsides by a Saturn 5 at blastoff than a pair of Pratt and Whitneys. And during the launch at least, I have a better view of the stars from a DC-3.

Four minutes later the shadowed land slips from beneath us and we are over the darker ocean. It is, as the gooney bird flies, 2,091 miles to Honolulu. The tower bade us farewell with a hint of good riddance, and George Washington has switched to the en route radio frequency.

As I ease *Savaii* upward, four thousand memories assail me, for I have as many hours in DC-3s. In cramped cocoons nearly identical to this one, I had frozen in the arctic and melted in the tropics. I had been sublimely content in autumn evenings above the shores of Lake Erie and awed by the aeronautical cruelties lurking in Catskill thunderstorms. Living so many hours in these noisy little cavities, I had belly-laughed over inconsequentials, dreamed ambitions never to be satisfied, scribbled naive notes for books I would never write, made lifelong friends, and wept for some who were slain. In these drafty little cubicles of aluminum lined with green leather I had known shame, lust, triumph, and near despair. And I had learned humility.

It is little wonder that after an absence of nineteen years I have absolutely no difficulty reaching for every control, absorbing the information offered by every instrument, or responding to the tolerant flight demands of a DC-3. These things, all of them, are engraved in my mind forever. Like the prisoner of Zenda, I know my cell.

We are supposed to report our arrival over "Briney," a radial intersection twenty-two miles offshore. We have not troubled to inform Air Traffic Control that we lack the electronic gear for such an exact fix. They would not understand our reasoning or our temporary reliance on dead reckoning. Yet air traffic controllers, like all the rest of us, are comfortable with the familiar. For years they have been clearing jets to "Briney" and no matter what their computers tell them they obviously cannot believe the near static target on their radar screens. Are we a balloon? How can we be so lackadaisical in reaching "Briney"?

They call three times to ask when we estimate arrival. When George Washington gives them an educated guess they wait only a few minutes before calling again. Wouldn't any self-respecting flying machine long ago have passed "Briney"? In contrast to the other blips swimming quickly about their screens we are apparently stuck in the celestial mud.

"They won't believe me!" George Washington's eyes are hurt. I wonder what would happen if I should pick up my own microphone and scold them for insulting his name.

To the controller's relief we eventually decide we are arrived at "Briney." He is rid of us. We have also struggled to 6,000 feet. The moon peers benignly over my left shoulder as I level *Savaii* and ease the engines into long-range cruising power. Far above us moves the modern world of flight. Here, with our two engines snoring like contented pigs we slide along smoothly enough. Anachronism be damned! The top-gallants and royals are set. The breeze is drawing fine. Sail on!

Soon we are free of our radar fetters, and George Washington retires to his small navigating table situated just behind my seat. A curtain between us shields his light from the cockpit, but where it should button against the curving side of the fuselage there is a separation. The buttons are missing and so a narrow band of reflected light is created on my side window. As if observing him on a miniature television screen I can watch George Washington settle down to work at his flight log and chart. He opens a book of tables and scribbles with his pencil. He wets his lips several times and frowns. He is making computations concerning the stars and planets which will be our beacons during the balance of the night. Suddenly, I am sorry for those who no longer use the heavens to guide their way.

You there, aloft in your jets so high above us! Are you content with the magic of your Inertial Navigation System? Do those impersonal, ultra-efficient, cold green numbers flicking across the panels of your obsequious machine now seem to match the beauty of the stars we use in our subterranean world? If so, I fear you are lost men bound to genuflect before an electronic marvel and I do not envy you.

Now for us, there is only the firmament, the vast ocean, and ourselves.

I am reluctant to turn on the automatic pilot, wishing to prolong this very special, rather sensuous experience, the return to an old and willing love. *Savaii* responds to my slightest touch . . . a change of altitude twenty feet . . . a few degrees off course . . . ah. Beyond the windshield is *Savaii's* broad snout. The moon outlines it clearly now and it droops down to a line of fluff balls, innocent little clouds marching along the black line of the sea.

For a time I seem to be alone with *Savaii*, staring at the fluorescent instruments exactly as I had done through so many long nights, almost hypnotized by their somnolent gentle motions, slipping pleasantly into that unique trance peculiar to night flying, that strange mixture of alertness and lethargy which somehow magically adds up to inner peace.

How many nights had I sat in just this way? During the Korean War it seemed we carved a track through these same skies; so many times did we pass back and forth with our cargos of fresh men out, and torn men home. And before that there was a steamship company which employed me to fly their first attempts to leave the surface for the air. But those flights were made in much bigger four-engined airplanes, so heavily manned and relatively sumptuous there were two bunks for resting when we pleased.

Dodie is standing in the moonlight beside me. I pretend to be working instead of luxuriating. She blinks at the rows of instruments.

"Coffee . . . tea . . . or milk?" Her nasal tone is part of our agreement. I had said that she could come if she rehearsed the stewardesses' chant to perfection.

"No thanks. Sit down and fly."

She needs a hundred-mile cross country for her private license. Though flying *Savaii* will not satisfy the FAA it will at least make a startling entry in her logbook. I take my hands and feet away from the controls. *Savaii* wavers momentarily, then settles obediently back to business. In the moonlight I watch Dodie's knuckles turning white. It will pass. *Savaii* is far larger and heavier than the little planes she knows. She will soon relax.

John Best has come forward. He scratches at the curly locks of his hair and I wonder if all New Zealanders are curly-headed since George Washington's hair is much the same.

"Are you ready to go on the fuselage tanks?"

"Wait. Let's burn off the auxiliaries another fifteen minutes."

"Right."

"Did you like the United States?"

"Yes . . ."

I wonder at a strange lack of enthusiasm in his voice.

"You wouldn't care to migrate . . . become a citizen?"

"No, thank you, skipper." Cold and flat. Too bad. John Best is the kind of young man we need.

Midnight. The cockpit has been like a miniature stage upon which our limited dramatis personae appear, speak their few lines in the subdued light, then exit into the darkness from whence they came.

Dodie has gone back to the cabin to tinker with the buffet. The heater is not working properly. John Best has turned four fuel valves so that now *Savaii*'s engines are sucking life from the long metal tanks which are lashed in the area normally occupied by passenger seats. Each tank holds four hundred gallons and passage between them is barely possible.

Two hours ago George Washington came forward to announce that his first star fix placed us in central China. Then he chuckled and pounded on the side of his head.

"I not only mixed up my Greenwich time, but was looking in the south latitude tables instead of north."

Later he returned with a second fix which was as near perfect as man could ask.

George Washington is a good navigator. Now, reflected in my side window, I can see him winding his octant. There had been so many nights over the North Atlantic when I had performed the same manipulations although our octants were not nearly so fancy. In these same type airplanes, we had done our own navigating. While the copilot flew the airplane I would go back to the cabin, climbing over bodies and cargo with octant in hand. If lucky I could catch a significant star through one of the cabin windows or the narrow skylight above the toilet. Sometimes it was necessary to signal the copilot for a turn to the right or left and thus reveal a certain star. It was a clumsy arrangement and our "fixes," if they could be dignified with the title, were heavily dependent on imagination. And hope. But we made it to Labrador and Greenland and Iceland as we would make it this night to the island of Oahu. In the intervening years the only apparent change is the addition of an astrodome through which George Washington observes the identical stars.

Dawn. It comes slowly for we are bound away from it and even our leisurely pace delays our pursuit by the sun. We are weaving between towering cumulus which would be dwarfed to toadstools by high-flying jets but appear as formidable bastions to us. We pass in dreamlike sequence from one to another and around the next. We are a butterfly seeking its way through a forest.

George Washington is patiently trying to transmit a position report on the radio. He holds his flight log in one hand and the microphone in the other while he reads off the long list of numbers which describe our whereabouts, our future whereabouts, comment on the weather, and our fuel endurance. It is a frustrating business which must be attempted every hour. Listening to my own headset as we strive for the most elementary communication with our fellow men, I am disillusioned. As it was long ago. In an emergency we would be more dependent on God and Pratt and Whitney than electronics for our survival. In the midst of fantastic progress, aviation has neglected its Achilles heel. I cannot detect how en route long-range radio communications have made the slightest improvement in the past fifteen years.

Now the moon rides ahead of us. Against the pale sky of a new day it looks a fake, like something the property man forgot to take down. The stars have taken their leave so George Washington has retreated to the cabin for a well earned rest. He will resume his duties when the sun is high enough to offer a good shot. I have also rested, slumbering like a child while John Best twisted the knobs on the automatic pilot. And now it is his turn to close his eyes.

Dodie brings me tea, apologizing for its cool temperature.

"The buffet is cassé."

I tell her it is too early in the morning for such corny rhymes. She has also brought a sandwich of salami and cheese and while I munch at it and watch the glorious dawn, Dodie flies *Savaii*.

I glance down at the ocean and appreciate how the depths of the clouds are still wrapped in gloom. Yet a moment later, still munching, I see the left wing tip twinkle with the first touch of sun. Here is great contentment. All is as it should be in my aerial world.

Thirteen hours and fifty-three minutes after taking off from San Francisco, *Savaii's* tires kiss the runway at Honolulu. And again we have offended the normal order of things. The control tower, speaking its annoyed mind, confesses it knows not what to do with us. We wait, orphaned in the middle of the vast airport while great jets scream past. Who are we who dare bring ancient history into the hectic morning business of a great airport? Go away, flying Dutchman! There is no longer any appropriate nest for an aged DC-3.

Finally we are directed to a lonely tin hangar, itself an anachronism. A man from the State of Hawaii arrives in a yellow truck to collect $29 for the landing fee.

Twelve hours later we are airborne again, bound for Apia. And on this second night, climbing in the humid Hawaiian sky, I am suddenly possessed with the fancy that this whole flight is a dream. I will awaken any instant and this candy-floss airplane will vanish. My hands now caressing the controls will grasp only air. Or perhaps, as the lights of Honolulu sink into the depths, *Savaii* will become a submarine with this unbelieving Captain

Nemo trying to fit his anachronistic craft into reasonable harmony with the instructions rattling in my earphones.

Earphones? Anachronism upon anachronism. These heavy pre–World War II types always made me feel like a yoked ox.

Honolulu Departure Control has us on their radar screen. The controller himself is loquacious with local gossip. " . . . you have traffic at two o'clock."

We peer at the night and see a great nothing. Even the stars are obscured by a cloud level we have yet to reach.

"Traffic, slow moving . . . at ten o'clock . . . two miles."

There. The blinking lights of an aircraft off to our left. He slips swiftly overhead and is gone. In earlier days we might have been innocent of his presence.

The controller asks if we have "VOR" equipment.

"Negative."

I hope he will reply, "How quaint . . . " but there is only silence.

The omission of VOR in *Savaii* is deliberate. Of what use would such sophistication be when the simple islands of the South Seas have no stations to transmit the necessary signals? Our next radio navigational aid will be a plain old-fashioned, nondirectional beacon located on Canton Island, over 1,600 miles of ocean and sky to the southeast. There is nothing in between. Now the situation is the same as our departure from San Francisco. Lacking the sophisticated electronic aid of VOR equipment we cannot depart the busy Honolulu area with sufficient exactitude to please Air Traffic Control. We are quite capable of wandering off on our own, but such carefree license was for leather-jacketed country bumpkins—not for the now in aviation.

"I will vector you on course," the controller announces. "Turn right to one eight five degrees."

I oblige. Big Brother is watching us. He can see aircraft we cannot see so it is wiser to let him escort us toward the great outdoors. Where we are bound the entire sky will more certainly be ours.

"Turn left to one seven zero degrees. You are now forty miles southeast of Koko Head . . . good night."

Thank you and good night, dear Big Brother. You cannot imagine how silly it feels flopping around in your radar controlled world on the beak of this ancient pelican. We are not pressurized so we can open the side windows when we please and toss our gum out or the wrapping of a sandwich or anything else which is messy and displeases us. We can open the window and stick out our noses and sniff at the moist lukewarm air, or peer down at the black sea, or enjoy an unadulterated view of the heavens. This, Big Brother, is something you cannot do in a jet.

Midnight again. At this time last night we were far to the east of Honolulu in regions heavily traveled by sea and air; here with our chances of seeing another aircraft or any ship on the sea infinitesimal, I rediscover a wonderful loneliness. Once, the New Bedford whalers sailed this area, and long before the incredible migration of Polynesian peoples followed this same general line in reverse until they came upon the Hawaiian Islands. And during World War II the skies were busy. But now . . . nothing. Our first flights across the North Atlantic had been much the same. We were entirely self-reliant, a condition which sweetens all loneliness. We worried about many things, but not about collision.

We are cruising at 6,000 feet again. The stage moon has been hung for another performance, this time shoved around until it is perched on *Savaii's* nose. The cumulus buildups which normally surround the Hawaiian land masses have been left behind. Below there is only the sea shimmering in the moonlight with here and there lost dumplings of vapor looking for a parent.

It is dark in the passageway which leads aft from the cockpit. George Washington is standing on a ten gallon oil drum taking his first fix of the night. The octant hangs from the center of the star-studded astrodome and seems to be a projection of George Washington's body.

John Best has switched to the cabin-ferry tanks and is standing beside me. His youthful face is intent on the fuel pressure gauges. He wants to be certain there is no air lock in the ferry system. In this vigilance he has my hearty endorsement.

"She seems to be feeding fine, skipper."

"Yes . . ."

John Best reaches for the booster pump switches and flips them off one at a time. The pressure needles sag and then revive. He turns back into the blackness and soon Dodie arrives to sing the coffee-tea-or-milk song. There is a look in her eyes which has nothing to do with refreshment.

"You really came up here to fly, didn't you?"

"I confess . . ."

"Take over, Mrs. Mitty. The sky is all yours."

For a time I sit half-dreaming in the square of moonlight framed by the windshield. And as inspired by so many similar nights I find myself marveling at fortune's inexplicable arrangements. During those easily memorable times when I flew open cockpit Wacos, Birds, Ryans, Stearmans, and anything else I could beg or borrow, how could I have envisioned these circumstances? Below is an ocean with the nearest land already hundreds of miles away. On my right is a young lady flying a heavy twin-engined aircraft better than I had hitherto thought she might do. Yes, perhaps I should stow my male chauvinist helmet and goggles away forever and admit there are no longer any great physical demands upon a pilot. At least it was some consolation to realize "they" have not so far automated the weather.

Or man's resistance to fright—as I would later be reminded.

Hours later it is my turn to rest in the cabin. I lie down on our makeshift bunk and discover the belt of Orion framed in the nearest window. There is Rigel on the right, Betelgeuse on the left, and farther out in left field is Procyon twinkling as brightly as I have ever seen it. Here, near the tail, the engines' muffled drone is soothing and mixed with the hissings of countless drafts spewing from as many small openings in the fuselage. It is true non-supercharged fresh air and I breathe deeply of it. Who could not sleep here with the tail swinging so gently back and forth as if some long gone aviator would rock me in the cradle of his heights?

Just before dawn I make my way forward between the fuel tanks and pause by George Washington's tiny cubicle. He has crossed the last of three small lines and holds the tip of his pencil upon it.

"I doubt if I'll have another fix until mid-morning. We are coming up on the intertropical front now and I should suppose we'll do a bit of bouncing about."

We have passed the equator and are in a region once known to sailing ship men as the Doldrums.

"Why don't you get some sleep?"

"I rather think I shall. I'm tired of bucking the bloody radio. Nandi is guarding us now."

Nandi is in Fiji, a long way over the horizon—like 2,000 miles.

I move forward to the darkened cockpit and tap John Best on the shoulder. He surrenders his seat with a smile.

"Just in time. There's some fire up ahead."

During my absence the moon has rolled from the left side of *Savaii's* nose to the right. Presently it occupies a lozenge-shaped clear space in the sky and illuminates a long wall of cumulus which extends much higher than *Savaii* could ever climb. Occasionally the fat and bulbus tumors, charged with their interior lightening, flicker brilliantly. Then they become quiescent for a while and soon, as if commanded by an energy-hungry Wotan, commence flashing again.

I watch the show from my front-row balcony seat. It appears to be no more of a threat than any other intertropical front I have ever seen, which if reviews were given for spectacle would place it considerably below a line of Appalachian thunderstorms and far down the honors list from the permanent thunderstorm front which lies off the west coast of Africa. There I had played audience to extravaganzas so visually terrifying I had wanted to crawl under my seat and hide, yet once a part of the performance I discovered they invariably had more bravado than bite. But this?

I remember a night when I had been overly casual about entering a line of thunderheads assembled over the Irish Sea and in two minutes found myself praying audibly for immediate salvation. Obviously my airplane would not survive. I resolved then and there that I would never trust a line of cumulus no matter how flabby their appearance.

Dodie, now awakened, joins me in the cockpit while I guide *Savaii* along winding cloud streets pressed on both sides by gigantic ramparts. Occasionally we reach a dead end and plunge into cloud. Then rain hisses at the windshields, *Savaii* is rudely jostled, and I ask Dodie to pull on a bit of carburetor heat. But these sessions are brief, not at all ugly, and soon we are gliding along another street. There are times when it is necessary to change course as much as twenty degrees to avoid the larger anvil-headed cumulus which really might be grisly. I always compensate a like number of degrees in the opposite direction lest George Washington have complaint that his carefully plotted course has been fouled. Our extra wanderings will not consume more than ten minutes. Why trade this much peace for that much war?

After an hour or so, when the dawn light has overpowered the moon's, we pass beyond the front. Now there are strange rills of curiosity passing through each of us, for soon there must be something to see. There? No . . . it is too early. Another ten minutes . . . maybe fifteen.

There!

Canton Island is an ill-defined bronze blob on the sea horizon which appears to have been dropped from the cloud stretching above it. All of *Savaii's* crew are gathered in the cockpit for their first sight of land since the night before. The intervening time has been like a passage from another life.

Very abruptly we are brought back to our immediate existence.

For suddenly there is a violent spasm of shudderings and regurgitations from both engines.

Our human responses are immediate. Dodie's hands try to squeeze juice out of the plastic control wheel. George Washington, ever the cool one, purses his lips and frowns. John Best disappears into the cabin.

With the speed of a d'Artagnan I turn on both main fuel tanks and hit the booster switches. When the engines settle back into their usual sonorous melody and my heart slides back down and off my tongue, I find it difficult to believe a former helmet and goggle man could have moved so fast.

John Best returns to the cockpit wearing the frown of a man betrayed.

"I don't understand it. The cabin tanks were still reading thirty gallons on my measuring stick."

"Do you abide by Murphy's law in New Zealand?"

"Ah? . . . Yes."

"When the plumbing system was installed someone bent a pipe up when it should have been bent down, or vice versa."

"It must be."

Fright, being merely a spark of fear, is much more quickly extinguished. Having swallowed my heart I affect a nonchalance suitable to the occasion. For lack of other gestures to fix the image I calmly call Canton and give our identification and destination.

At once I regret having opened the liaison. For the man who resides somewhere on that lonely atoll asks, "Are you landing Canton?" There is an unmistakable plaintiveness in his voice—a yearning.

"Negative."

From the atoll there is only the silence of disappointment. Then hopefully, should we change our minds, the voice recites the surface wind and altimeter.

"Thank you . . . so long." Even a little DC-3 has managed to pass over Canton Island.

Once we leave Canton it seems we have already arrived, although Samoa lies almost seven hundred nautical miles beyond. Yet it is close enough to rouse the homing instinct in half our crew. George Washington goes about his navigating with increasing zest instead of succumbing to the natural wilting which marks the very last part of most long flights. He keeps trying to tell Nandi our estimated arrival time in Apia, though he might as well holler down a barrel which is exactly what both he and the Nandi station operator sound like.

I turn to chide him. "You've been homesick."

"Of course."

John Best is also becoming increasingly restless. There is no room to pace, but he makes innumerable trips from the cabin to the cockpit. He will stand beside me for minutes, looking over the nose, scanning the horizon although he knows very well it will be hours before any land will appear. He borrows my pocket computer so he can calculate our true airspeed.

"One hundred and sixty knots true. . . ." Not quite true because his heart has added a knot for hope and perhaps two more knots for anticipation and his heart has hinted to his eyes that if he leans far enough to one side when he reads the airspeed indicator it will appear to read faster than if he looks directly at it.

Now suddenly, in a frenzy of activity, everyone is taking pictures of each other and of the sky which is now dulled by a high overcast and certainly the least photogenic sky since we left San Francisco. I put it down to a mutual urge to disperse minutes, something to do while the faithful engines drone on, something to keep hands occupied while time segments drag between arithmetical fact and desire.

In contrast I become morose and find myself behaving like a sentimental fool. In spite of a certain weariness I do not really want this anachronistic flight to end. I touch the elevator trim tab in a way that should be reserved for a woman, then catch myself pretending to wipe at a spot of grease. I know it will be a very long time before I will fly a DC-3 again.

Faleolo is the airport for Western Samoa, located a long twenty-two miles from Apia. It is one of the world's fast vanishing grass airports, which makes it a sensuous pleasure to land any airplane.

It is a strip of green confined by the sea on one side and ranks of coconut palms on the other. George Washington had given our arrival time hours before. He had predicted 1:35. It is 1:35.

I point to the brooding island, most of which is shrouded in gray, soggy-looking clouds.

"You lied, George Washington. We will not be on the ground until thirty-eight."

Three minutes. Three minutes deviation in a flight that had commenced over 4,000 miles away.

I call for the landing gear to be lowered. George Washington recites the litany of the cockpit check list, and after fifteen hours in the sky we slide down through the warmth to the grass.

There are many people waiting to greet us. They place welcoming leis around our necks and admire *Savaii* as if she were the sleekest supersonic transport.

Walking away from *Savaii* I paused to look back at her. And I saw beyond the people clustered all around her, volubly expressing their wonder. Suddenly I knew my long nourished envy of astronauts had been eased. For had our aerial voyage been beyond the moon, I thought, our sense of detachment would have been much the same. Now I knew there had been no true measurement of the distance from our fellow earthlings. *Savaii*, a phantom from the past, had been our space ship carrying us above reality and in her we had made a reluctant reentry. Thus may the useful life of a thing be prolonged and that of some men temporarily exalted.

Dossier on Principal Characters

(IN ORDER OF THEIR APPEARANCE)

ARMSTRONG-WHITWORTH ARGOSY

Span . . . . 90 feet
Power . . . . 3 Armstrong-Siddely Jaguar III, 385 hp. each
Passengers 18–20
Gross weight . . . . . . 18,000 pounds
Range . . . . 415 miles
Cruising speed . . . . . 90 mph.
Landing speed . . . . . 50–55 mph.

Phased out by 1933

DE HAVILLAND (DH-4)

Span . . . . 42 feet, 5½ inches
Length . . . . 29 feet, 11 inches
Gross . . . . 4,297 pounds

Top speed . . . . 124 mph. (only attempted by the military)
Cruising speed . . . . . 100–105 mph.
Power . . . . Liberty 400–420 hp. A 12-cylinder engine with "balls." Pilots who sat behind them never forgot their formidable
rumble. Occasionally cursed with plumbing problems.

Approximately 5,000 were built by the end of World War I.

FAIRCHILD 71

Wingspan . . . . 50 feet, 2 inches
Gross weight . . . . 5,200 pounds
Useful load . . . . 2,500 pounds
Payload . . . . 1,427 pounds (as land plane)
Maximum speed . . . 138 mph.
Cruising speed . . . . . 110 mph. (with optimist at controls)
Cruising speed . . . . . 100 mph. (equipped with floats and same optimist)
Landing speed . . . . . 57 mph.
Rate of climb . . . . 900 fpm. (if not loaded by bush pilot)
Ceiling . . . . 15,500 (and has done better when not working for a living)

The single-engined Fairchild series were originally flown
by Colonial Airways on AM 1 (Air mail route One). The
last known Fairchild still operational (1972) is owned by
Captain Herbert Harkom of American Airlines who did
his own restoration.

BOEING 40-B-4

Span . . . . 44 feet, 2 inches
Length . . . . 33 feet, 3 inches
Gross weight . . . . 6,075 pounds
Empty weight . . . . . . 3,722 pounds
Cruising speed . . . . . 125 mph. (according to manual)

Cruising speed . . . . . 110–115 mph. (according to fact)
Range . . . . 500 miles plus a few
Service ceiling . . . . . 16,100 feet (according to manual)
Service ceiling . . . . . 13,500 feet (according to fact)
Power . . . . Pratt and Whitney Hornet, 525 hp.

GRAF ZEPPELIN

Length . . . . 775 feet
Diameter . . . . 100 feet
Gas volume . . . . 3,037,000 cubic feet hydrogen
Fuel . . . . 918,000 cubic feet "Blau" gas (similar Propane)
Power . . . . 5 Maybach, 520 hp. each
Passengers . . . . 20 for long day and night hauls
Cruising speed . . . . . 76 mph. (average)

Graf Zeppelin flew a total of 1,060,000 miles in 16,000 flying hours. Crossed South Atlantic 140 times, North Atlantic 7 times.

CURTISS WRIGHT CONDOR*

Span . . . . 82 feet
Length . . . . 48 feet, 10 inches
Gross weight . . . . 17,500 pounds
Useful load . . . . 6,035 pounds
Top speed . . . . 171 mph.
Cruising speed . . . . . 145 mph.
Landing speed . . . . . 62 mph.
Initial climb . . . . 850 fpm.
Range . . . . 560 miles
Power . . . . 2 R-1820-F Wright Cyclones, 700 hp. each

* Condors were flying for China National as late as 1941. Others were flying in Costa Rica the same year.

POTEZ 25A

Span . . . . 46 feet, 7 inches
Length . . . . 30 feet, 2 inches
Empty weight . . . . . . 2,601 pounds
Gross weight . . . . . . 4,338 pounds
Maximum speed . . . 136 mph.
Cruising speed . . . . . 105 mph.
Range . . . . 310 mph.

Wooden construction, fabric-covered wings, plywood fuse-lage. Essentially a military aircraft.

BREGUET 14

Span . . . . 47 feet, 1¼ inches
Length . . . . 29 feet, 6 inches
Empty weight . . . . . . 2,729 pounds
Gross weight . . . . 4,374 pounds
Cruising speed . . . . . 77 mph.
Landing speed . . . . . 43 mph.
Range . . . . 285 miles
Power . . . . Various—Renault 300 hp. or Lorraine 375 hp. Also Fiat, Hispano Suiza, and even a few Libertys.

Built in 17 different versions.

SHORT S-17 KENT (known as SCIPIOS)

Cruising speed . . . . . 105 mph.
Span . . . . 113 feet
Length . . . . 78 feet
Range . . . . 4 hours plus a little for reserve
Power . . . . 4 Bristol Jupiter XFBM, 555 hp. each
Gross weight . . . . . . 32,000 pounds

JUNKERS — 52/3M

Span . . . . 95 feet, 11½ inches
Empty weight . . . . . . 11,785 pounds
Gross weight . . . . 20,282 pounds
Capacity . . . . 15–17 passengers plus 3 or 4 in crew
Cruising speed . . . . . 152 mph. (in a clean airplane which has not suffered too
                                  much abuse)
Landing speed . . . . . 62 mph. (slower in a power stall)
Take-off . . . . 1,115 feet
Ceiling . . . . 17,060 feet
Range . . . . 568 miles (wise men thought of it as 500)
Power . . . . 3 600 hp. BMW Hornets or 725 hp. Hornets or Pratt and
                                  Whitney S1E-Gs. A few were equipped with Junkers diesel
                                  engines.

FORD TRIMOTOR 5-AT-B

Length . . . . 49 feet, 10 inches
Span . . . . 77 feet, 10 inches
Empty weight . . . . . . 7,576 pounds
Gross weight . . . . 12,576–13,250 pounds, depending on cabin arrangements
Payload . . . . 2,720–3,044 pounds, depending on cabin arrangements
Maximum speed . . . 142 mph.
Cruising speed . . . . . 122 mph. (according to Ford claims)
Actual cruise . . . . . . 105–110 mph. (according to realities)
Landing speed . . . . . 65 mph.
Ceiling . . . . 18,500 feet
Range . . . . 540–610 miles
Power . . . . 3 P & W Wasp, 420–450 hp. each

DOUGLAS DC-2

Passengers . . . . 14 in American version; 8–11 KLM version
Span . . . . 85 feet
Length . . . . 61 feet

Gross weight ...... 18,200 pounds
Empty weight ...... 12,010 pounds
Alleged cruise ..... 200 mph.
Actual cruise .... 175–180 mph.
Landing speed ..... Approximately 65 mph. (pilots had different opinions
                                    according to their style)
Range .... 1,200 miles plus a few
Special virtue ...... 23,600 feet
Power .... Wright F3A, 710 hp. (Cyclone)
Service ceiling ..... Could successfully carry more ice than any contemporary.

DOUGLAS DC-3 (DST)

Maximum speed ... 212 mph.
Cruising speed ..... 180 mph. (honest)
Landing speed ..... 65 mph. (Honest. Better keep 85 "over the fence")
Ceiling .... 23,100 feet
Passengers .... 21 (in later versions, 28)
Engines .... Depending on airline choice. Wright G-102, or Pratt and
                    Whitney 1830s.

BOEING 314 (CLIPPER)

Span .... 152 feet
Length .... 106 feet
Passengers .... 68 day, 38 in berth accommodations; 89 in nonluxury
                    wartime versions.
Gross weight ...... 84,000 pounds
Stall speed .... 63 knots power on, flaps down, to 78 knots, power off and
                    flaps up.
Top speed .... 199 mph.
Cruising speed ..... 184 mph. (in practice more like 180)
Range .... 5,200 miles
Engines .... 4 Wright Cyclones, 1,500 hp. each. (Later 1,600 hp.)

DE HAVILLAND MOSQUITO
Mark Numbers to 35

Top speed . . . . .Depending on Mark number, from 375 mph. to 410 mph.

Stall speed . . . . . . .Flaps down and 17,000 lobs. 103 mph.

Range . . . . .Approximately 1,200 miles depending on Mark number and type.

Op. Altitudes . . . . . .To 40,000 feet in later Mark numbers with pressurized cabins.

Power . . . . .2 Rolls-Royce Merlin. Variable according to Mark number and supercharger gear in use. To 2,600 hp. each.

Twin-engine all purpose aircraft of wooden construction.

DOUGLAS DC-4 (C-54, R4-D)

Span . . . . .117 feet

Length . . . . .93 feet, 10 inches

Weight empty . . . . .37,250 pounds

Gross weight . . . . . . .65,000 (sometimes loaded to 73,000+!)

Landing speed . . . . 110 mph. ("over the fence"). Touch down about 90–100 depending on load.

Cruising speed . . . . .Approximately 190–200 mph. depending on load.

Power . . . . .Pratt and Whitney Wasp R-2000-7 (11 and 13s)

Rare characteristic for any airplane: Stall speeds were better (lower) with power off than with power on—loads being equal.

BOEING 377 STRATOCRUISER

Length . . . . .110 feet, 4 inches
Span . . . . 141 feet, 3 inches
Cruising speed . . . . .300–340 mph. at 25,000 feet
Stall speed . . . . .93 mph. sea level, 110,000 pounds
Max. range . . . . .3,000 miles, no reserve, 24,000 pounds payload.
Max. takeoff load . . .145,000 pounds
Max. payload . . . . . .30,000 pounds
Empty weight . . . . . .85,000 pounds
Power . . . . .4 Pratt and Whitney Wasp Major R-4360, 3,500 hp. with water injection.

Pressurized cabin equals 5,500 feet at 25,000 feet.

Accommodations for passengers varied considerably according to service and airline, with 75 about the average.

MINOR CHARACTERS

Pitcairn Super Mailwing

Span . . . . .35 feet (upper)
Span . . . . .32 feet, 1 inch (lower)
Length . . . . . . .24 feet, 10 inches
Empty weight. . 2,294 pounds
Gross . . . . .4,000 pounds
Max. speed . . .145 mph.
Cruise . . . . .122 mph.
Landing . . . . . .60 mph.
Ceiling . . . . . . .16,000 feet
Range . . . . .600 miles
Power . . . . .Wright J-6, 300 hp.
Price $12,500

Boeing Monomail

Span . . . . .59 feet, 2 inches
Length . . . . . . .41 feet, 2 inches
Empty weight. . 4,626 pounds
Gross . . . . .8,000 pounds
Max. speed . . .158 mph.
Cruise . . . . .137 mph.
Landing . . . . . .57 mph.
Ceiling . . . . . . .14,700 feet
Range . . . . .540 miles

Northrop Alpha

Span . . . . 41 feet, 10 inches
Length . . . . 28 feet, 5 inches
Empty weight . . 2,679 pounds
Gross . . . . 4,500 pounds
Max. speed . . . 170 mph.
Cruise . . . . 145 mph.
Landing . . . . . . 60 mph.
Ceiling . . . . 19,300 feet
Power . . . . Wasp R-1340-C, 420 hp
Price $21,500

Blohm und Voss HA 139

Span . . . . 88 feet, 7 inches
Length . . . . 63 feet, 11¾ inches
Empty weight . . 22,839 pounds
Loaded . . . . 38,581 pounds (for
                     catapult launch)
Cruise . . . . 161 mph.
Ceiling . . . . 11,482 feet
Max. range . . . 3,293 miles
Power . . . . 4 605 hp. each,
                     Junkers Jumo diesel
                     engines

Consolidated Commodore "16"

Span . . . . 100 feet
Length . . . . 61 feet, 8 inches
Empty weight . . 10,550 pounds
Gross . . . . 17,600 pounds
Capacity . . . . . 10 passengers, 600
                    pounds baggage and
                    cargo, or 18 passen-
                    gers and 485 pounds.
                    Crew of 3.
Max. speed . . 128 mph.
Cruise . . . . 108 mph.
Landing . . . . . . 60 mph.
Range . . . . 1,000 miles
Price $125,000

Savoia Marchetti S-73

Span . . . . 78 feet, 9 inches
Length . . . . 57 feet, 3 inches
Empty weight . . 15,278 pounds
Gross weight . . 22,993 pounds
Cruise . . . . 173 mph.
Landing . . . . . . 56 mph.
Ceiling . . . . 24,278 feet
Range . . . . 633 at cruising pow-
                     er. Maximum 994
                     miles.
Power . . . . . . .3 Piaggio Stella X.
                    RC. (Also equipped
                    with Wright Cyclone
                    1820 engines or 800
                    hp. Alfa Romeos.)

Farman F-3X (Stork)

Span .... 62 feet, 4 inches
Length .... 44 feet, 10½ inches
Empty weight . . 6,613 pounds
Gross .... 11,023 pounds
Cruise .... 108 mph.
Ceiling .... 13,000 feet
Range .... 403 miles

Fokker Super-Universal

Span .... 50 feet, 8 inches
Length .... 36 feet, 7 inches
Empty weight . . 3,000 pounds
Gross .... 5,150 pounds
Max. speed ... 138 mph.
Cruise .... 118 mph.
Landing ...... 56 mph.
Ceiling .... 18,000 feet
Range .... 675 miles
Price $17,500

Lockheed Orion

Span .... 42 feet, 9¼ inches
Length .... 28 feet, 4 inches
Empty weight . . 3,640 pounds
Gross weight . . 5,400 pounds
Top speed .... 226 mph.
Cruise .... 195 mph.
Landing ...... 65 mph.
Power .... P & W Wasps, 450
hp. Later models
with 550 hp.
Cost .... $25,000 in 1931
$20,000 in 1934

DH-86

Span .... 64 feet, 6 inches
Gross weight . . 9,200 pounds
Capacity ..... 10–12 passengers
Cruise .... 145 mph.
Range .... 450 miles
Power .... 4 200 hp. De Havil-
land Gypsy Six
engines.

Piaggio (L1 and L2) Royal Gull

Span .... 44 feet, 4¾ inches
Length .... 35 feet, 5¼ inches
Empty weight ...... 4,682 pounds
Gross weight ...... 5,996 pounds
L-1

Cruising . . . . 169 mph., as advertised

Cruising . . . . 158 mph., as the truth flies

L-2

Cruising . . . . 192 mph., as advertised

Cruising . . . . 178 mph., with enough fuel to fly beyond the horizon

Landing . . . . 72 mph., could be somewhat less with full flaps and careful handling. Stalls very sharply.

Range . . . . 900 miles. The last 100 could become very anxious.

Power (L-1) . . . . 2 Lycoming 270 hp. geared engines.

Power (L-2) . . . . 2 Lycoming 340 hp. geared and supercharged.

Index